T0261218

DAVID S. CAUDILL

WITH A FOREWORD
BY HARRY COLLINS

EXPERTISE IN CRISIS

The Ideological Contours of Public
Scientific Controversies

BRISTOL
UNIVERSITY
PRESS

First published in Great Britain in 2023 by

Bristol University Press
University of Bristol
1–9 Old Park Hill
Bristol
BS2 8BB
UK
t: +44 (0)117 374 6645
e: bup-info@bristol.ac.uk

Details of international sales and distribution partners are available at
bristoluniversitypress.co.uk

British Library Cataloguing in Publication Data
A catalogue record for this book is available from the British Library

ISBN 978-1-5292-3091-8 hardcover
ISBN 978-1-5292-3092-5 ePub
ISBN 978-1-5292-3093-2 ePdf

Cover design: Lyn Davies Design
Front cover image: Alamy/ART Collection
Bristol University Press uses environmentally responsible
print partners.
Printed and bound in Great Britain by CPI Group (UK) Ltd,
Croydon, CR0 4YY

For Marilyn

Contents

About the Author

David S. Caudill is Professor and Arthur M. Goldberg Family Chair in Law at Villanova University Charles Widger School of Law in Villanova, Pennsylvania, and Senior Fellow on the University of Melbourne Law Faculty. He holds a PhD in philosophy from the Vrije Universiteit Amsterdam, a JD from the University of Houston Law Center, and a BA in philosophy from Michigan State University. He previously taught at Washington and Lee University School of Law; prior to that appointment, he clerked for Judge John R. Brown in the US Court of Appeals for the Fifth Circuit, and practiced law in San Diego and Austin. He is the author of seven books and over 100 journal articles and book chapters in the fields of law and science, legal ethics, property law, and law and literature.

Acknowledgments

I am grateful to those who have offered comments on, and criticisms of, the analyses and arguments in this book, especially: (1) the anonymous referees in the proposal and manuscript submission phases of Bristol University Press's production process; (2) my fellow panelists and those who attended my presentation at the (virtual) Society for Social Studies of Science Annual Meeting, held at the University of Toronto, on October 6, 2021; (3) the participants, including Harry Collins and Rob Evans, in the (virtual) international workshop hosted by The Centre of the Study of Knowledge, Expertise and Science (KES), a research group based in the Cardiff University School of Social Sciences, Wales, on February 21, 2021; and (4) my colleagues at a Villanova Law Faculty workshop who responded to a presentation at an early stage in this project. Many of my contentions in this book were outlined in "Trust in science: the crisis of expertise as an ideological, and not only a scientific, controversy," published in the *Quinnipiac Law Review*, 40: 237–87 (2022). Some of the materials on Bruno Latour were presented on December 9, 2019, in a keynote address to the 2019 Workshop on Law, Technology, and Humans, Queensland University of Technology, Brisbane, Australia, and published as "Expertise in political contexts: Latour *avec* the Third Wave in science and technology studies," *Law, Technology, and Humans*, 2(2): 4–21 (November 2020). Some of the materials on Dutch Golden Age church interiors were published in "Emanuel de Witte's *Interior of the Oude Kerk, Delft*: images of life as religion, individualism, and the critique of legal ideology," *INDEX Journal*, 2 (Symposium on Law and Art) (2020). I am grateful to the editors and peer reviewers of these journals for their assistance and advice. Thanks to Dr Darrin Durant, University of Melbourne, for his guidance with respect to Wittgenstein's philosophy, and to Dr Martin Weinel, Cardiff University, for his helpful insights on counterfeit (as opposed to genuine) scientific controversies. Thanks also for the summer 2022 research grant to complete this book provided by Villanova University Charles Widger School of Law. Finally, I want to thank Paul Stevens, Publisher and Head of Interdisciplinarity and Digital, and Georgina Bolwell, Senior Editorial Assistant, both at Bristol University Press, for their discernment and generous spirit throughout the publication process.

Foreword

Not many people know that before he became an academic, Dave Caudill would sometimes sit at the end of a runway in Germany in an F-4 Phantom jet with an armed nuclear weapon, waiting for the alarm that would send him to release it on the forces of the Soviet Union. That was his earlier method of promoting peace. Now, the pen has taken the place of the sword. Like many of us, he sees the world getting ever nearer to mutually assured destruction as a result of the erosion of truth. A group of apprehensive people, including Caudill, think that a proper understanding and respect for science might help to stop the death of truth, followed by the death of all of us. However, this is not the kind of understanding of science that causes people to go around wearing "Trust the Science" T-shirts because we know science, most evidently science in formation, is too provisional and embedded in ordinary social life for that. Typically, the kind of science that confronts potential dictators and could limit their power to exercise "the will of the people" in any way they interpret it is the provisional sort of science that still could be wrong. We have to find a way to bring this kind of science to bear on public opinion without going back to slogans reminiscent of the 1950s. This is what some of us think of as the ambition of the Third Wave of science studies.

Too many social scientists take the easy way out, abandoning the old model so completely that science folds into public opinion; in this book, Caudill confronts the dilemmas. You cannot have checks and balances without elites. Caudill's treatment of this is original in a number of ways and unique in bringing his immediate post-military profession of lawyer into the debate. He knows how the paradoxes of expert witnesses confronting lay juries work out, and if nothing else, this aspect of the book will make it indispensable.

Harry Collins
Cardiff, 2022

Preface

Within the so-called "culture wars" dividing many nations politically, there is a persistent controversy, intensified during the COVID-19 pandemic, over the trustworthiness of consensus science—the so-called "crisis of expertise." When the science concerning climate change, mask wearing, or vaccinations becomes politicized, it loses its mooring in scientific evidence and reduces the effectiveness of both regulatory laws and the voices of scientists. Perhaps counterintuitively, however, the solution is likely not to wear "Because Science" T-shirts, while insisting on "cold, hard facts" and diagnosing as stupid those who believe the scientific theories of marginalized, minority-view scientists. Indeed, a certain level of modesty—regarding the uncertainties and tentativeness of even the best science—is necessary for the type of understanding and communication that might convince someone to change their scientific beliefs. Unfortunately, the reaction of some scholars to the crisis of expertise is unwittingly to idealize consensus science by identifying an anti-science ideology in certain segments of the citizenry, while easily ignoring the ideological, almost religious, belief structures on both sides in the crisis of expertise. Indeed, the arrogance of those who believe in consensus scientists (a group to which I belong) probably increases distrust of experts. How do we reduce that conflict over consensus science? In Wittgenstein's formulation: "Conflict is dissipated in much the same way as the tension of a spring when you melt the mechanism (or dissolve it in nitric acid). This dissolution eliminates all tensions."[1] But what is the mechanism dividing us in scientific matters, and what can dissolve it?

The purpose of this book is to analyze the crisis of expertise in terms of ideology, by which I mean an inevitable worldview (and not the Marxian-inspired notion of a false consciousness, perhaps empowering a ruling class). I readily acknowledge the difficulty of useful discourse between groups who seem to live in different realities, and I draw on the recent work of sociologists who recommend modesty concerning probabilistic scientific models and data that can rarely be characterized as unchanging "cold, hard facts." I then propose that each side in the crisis of expertise be understood as quasi-religious believers in their facts, not in the sense of deistic belief, but rather as occupying an ideology or worldview. Finally, drawing upon Wittgenstein and those sociologists of science inspired by his later philosophy, I identify four types of experts in the crisis of expertise: (1) consensus scientists; (2) those who believe in consensus science; (3) marginalized scientists; and (4) those who hold marginalized scientific views. Expertise, that is, should not be associated with esoteric knowledge (or even correctness), but is rather the result of a community with shared practices and a common language. Reconceiving science as a field of numerous uncertainties, together with recognizing each side in the crisis of expertise as having faith-like commitments, will best serve the goals of self-understanding and persuasive communication with respect to scientific disputes in the culture wars generally and specifically in governmental policy contexts.

Introduction

We hear the questions so often nowadays from colleagues, friends, and family, whether in discussions of climate change,[1] the COVID-19 pandemic, or the safety of vaccines:[2] "How can those people ignore the obvious facts?"; "How can they be so lost in their bubble?"; and "Who are their so-called experts?" Of course, I do not mean to imply that it is only one side in the culture wars asking those questions; rather, both sides view the other as living inside a bubble or an echo chamber.[3] Fox News and CNN are "said [to] report as if from alternate universes."[4] These divisions have legal and policy consequences, as we have seen in the suggestion the Trump administration reflected an anti-scientific bias in appointments to head science-related government agencies, as well as in its response to the COVID-19 pandemic.[5] However, while scholars agree that 21st-century technological growth and the digital age has exacerbated the "tribal" divisions in the US and internationally,[6] the phenomenon of citizens living in "two different worlds," or in "alternative realities," is hardly new. A relatively random historical parallel—but one to which I will return as exemplary of our contemporary situation—is the sharp division between Catholics and Calvinists during the Protestant Reformation. Each side was convinced of both the righteousness of their cause—not only of their beliefs, but also of their acts of violence—and the dangerous blasphemy of the other.

More recently, about 30 years ago, Michiel Schwarz and Michael Thompson, focusing on risk assessment in policy contexts, highlighted the role of cultural cognition in ongoing clashes of contradictory certainties and plural rationalities.[7] Drawing on the myths of nature represented by some ecologists (nature as capricious, benign, perverse/tolerant, and ephemeral) and mapping them onto some anthropologists' representation of two dimensions of sociality (individual versus group and no external restrictions on choice versus external restrictions on choice) and four types of "rationalities" (fatalist, individualist, hierarchist, and egalitarian), Schwarz and Thompson identified four different orientations in technology assessment.[8] Thus, for example, the contradictory *certainties* held, respectively, by the producer of a genetically modified (GM) food product (that the product is safe) and an anti-GM activist (that the product is unsafe) can be explained by reference to differing perceptions of nature as, respectively, robust and vulnerable. That analysis is similar to another study in 2008 of cultural cognition in the debate over the risks of synthetic biology: politically conservative and religious opposition to synthetic biology is associated with an individualistic and hierarchical cultural worldview, whereas those with a communitarian and egalitarian worldview are less sensitive to the risks.[9] That study both confirmed the link between cultural values and disputes over environmental risks, and demonstrated a curious reversal of expectations:

> In general, individuals who hold relatively egalitarian values tend to be more risk sensitive and those who hold relatively hierarchical values more risk skeptical concerning technological and environmental risks. ... [R]ecognition of global warming and nuclear power risks, for example, tends to be associated with challenges to authority, [which] repels persons who are culturally hierarchical, politically conservative, and religious; recognition of synthetic biology risks, in contrast, coheres with resentment of a form of cultural secularism, symbolized by science, that is ... subversive of traditional forms of authority.[10]

Even the reversal, therefore, is explained in terms of cultural cognition. For Schwarz and Thompson, contradictory views of nature "lie beyond the reach of both orthodox (what are the facts?) scientific method and the conventional notion of 'decision making under uncertainty'":[11] "Another way of putting it is that each actor is perfectly rational, given his or her convictions as to how the world is. The situation is one of *plural rationality*."[12] Challenging the view that science, "for all its admitted uncertainties, is factual" and not driven by values, Schwarz and Thompson conclude:

> Anthropologists and sociologists of knowledge have shown us that what are considered facts depends ultimately on an accepted framework of social (and therefore evaluative) premises. Even scientific knowledge, whilst not perhaps wholly fluid, is certainly plastic in the sense that it is socially negotiated (science being a social activity) and molded by values of various kinds.[13]

The relevance of Schwarz and Thompson's analytical framework to our current circumstances is that while much has changed, we should not be provincial and assume that our contemporary cultural divisions over scientific matters are new. On the other hand, the COVID-19 pandemic has certainly brought a new set of bright-line divisions, causing many to assert that they have never before seen this current level of distrust in consensus science.

A. Expertise in crisis

I focus in this book on what is today called the "crisis of expertise,"[14] which seems to be a subpart of the broader "culture wars," the latter of which might include the tribal divisions in the US (and many other countries) based on differing political parties, human values, economic priorities, and so forth. The "crisis of expertise," on the other hand, refers initially to the distrust of consensus science on the part of a movement or group of citizens, and, in most cases, those same citizens' strong belief in alternative, minority "scientific" views. This too appears to be a tribal division. The crisis also refers to the politicization of science, insofar as those who, for example, believe in "man-made" climate change (or in the efficacy of mask wearing during a pandemic), and those who do not, become associated with opposing political parties and even opposing politicians (who, like ordinary citizens, may or may not trust consensus science).[15] Recall Latour's recent observation: "[W]hether or not a statement is believed depends far less on its veracity than on the conditions of its 'construction'—that is, who is making it, to whom it's being addressed and from which institutions it emerges and is made visible."[16] There is distrust nowadays of the conventional institutions of scientific knowledge.

Numerous explanations of this phenomenon, and various proposed solutions, are discussed in this book, but my own focus is on two aspects of the crisis of expertise that are often overlooked in contemporary discourse. First, I argue that the ideological (or "quasi-religious") features of those occupying either side in the conflict over consensus science need to be acknowledged—examples of the failure to recognize belief structures on both sides of the controversy include "Trust Science, Not Morons" T-shirts, which accompany the claim that those who doubt consensus science are simply ignoring the "cold, hard facts" produced by science. By attending to this ideological aspect of the crisis of expertise, we can begin to emphasize the methodological and epistemological limits of consensus scientific knowledge, and enhance modesty on the part of those who believe in consensus science. Many sociological analysts of the crisis of expertise view such modesty as the key to communication and understanding between the two sides in the tribal controversies over science.

Second, I argue that we need to reconsider the nature of expertise, both as (1) a broader category that includes scientific *and* non-scientific experts, and as (2) a "practice" that does not necessarily align with, or signify, correctness or truth.

As I will explain, based on recent work in the sociology of scientific knowledge (SSK), a domain of expertise should be understood as simply an area of practice and language use shared by those in the domain, which can be as common as speaking English or playing tennis. As to the question of the correctness of expertise, I will suggest that under the foregoing definition of expertise, which turns out to be the most useful in analyzing the crisis of expertise, one can be an expert in witchcraft or astrology—there is a shared practice and a shared language among the followers of these bodies of knowledge, respectively, in (I presume) casting spells and reading horoscopes. Importantly, their expertise in no way validates the truth of witchcraft or astrology. In contrast to the conventional use of the term "crisis of expertise," wherein "expertise" is a reference to *consensus* scientists, the typology of expertise employed in this book breaks down the crisis into four groups of experts, namely: (1) the majority of *scientists* in a particular field who establish consensus; (2) the community of *citizens* who believe in consensus science; (3) the minority *scientists* who reject consensus science on the basis of alternative scientific models; and (4) the community of *citizens* who distrust consensus science and embrace minority scientific views. Each group employs a practice and a shared language, which results in a "form of life" (in Wittgenstein's terminology), or worldview. I rely on insights from the anthropology of religion to show how religious communities reflect *expert* practices and language—the manner in which the religious *live out* their beliefs reflects an expertise and, as Wittgenstein recognized, a form of life. Yet, in my formulation of ideological commitments, my reference to organized religion is only a demonstrative analogy, as ideology in the crisis of expertise does not imply sacred rituals or belief in a deity.

Of course, the ultimate beliefs of those in Categories 1 and 2 is different from the ultimate beliefs of those is Categories 3 and 4. No group of experts, however, can rise above or escape having a set of beliefs that function as a worldview—there is no neutral ground in the crisis of expertise, even though each side frequently claims (wrongly) that its side is *not* ideological, but the other side is. I agree with those critics, discussed later, who argue that replacing that type of arrogance with some level of modesty, especially on the part of consensus scientists and their followers, is an important part of any solution to the distrust of consensus science by many US citizens. Moreover, as I hope to show, the call for modesty on the part of those who produce, and of those who believe in, consensus science finds helpful parallels in: (1) the identification of the cultural foundations of the scientific enterprise in the SSK; (2) Wittgenstein's notion of *Lebensform*; and even (3) 19th-century Dutch philosophical reactions to Enlightenment rationalism as ideological (long before the analogous postmodern rejection of the autonomous Cartesian subject), as reflected in Calvinist religious doctrine and Golden Age paintings in Holland. Readers may be surprised that I anchor my analysis of ideology (in the crisis of expertise), at least in part, in 17th-century art and 19th-century Dutch neo-Calvinist philosophy, but it is in those two contexts that the type of belief structures that we associate with "religious belief" suddenly lose most clearly their connection with a deity—everyone is in some sense "religious" and should be viewed as a believer with fundamental, faith-based commitments, whether based in religious texts *or* in the pretensions of Enlightenment rationality. The story of Dutch church interiors and neo-Calvinist critics of the French Revolution also exemplifies the sort of tribal divisions we experience nowadays and from which we need somehow to escape.

Notably, any solution to, or successful effort to tamp down the rhetoric within, the crisis of expertise may be different from the solutions (not addressed in this book) to the broader culture wars. It may not make sense to say that those who believe that Biden won the 2020 presidential election should be willing to entertain the possibility that he did not. However, in the field of scientific expertise, "cold, hard facts" are not so easily accessible. If each side believes that they have absolute truth, we are doomed to a discourse of calling one another stupid. It is never simply a matter of, in the words of Dr Fauci, "letting science speak."[17] Dr Fauci and Dr Birx may well both be correct that the Trump administration was anti-scientific, that President Trump would not listen to consensus science, and that Trump's advisors kept a parallel set of data[18] (like a fraudulent business with two sets of books!), but that does not mean that consensus science is a set of 100 per cent correct, eternal truths.

A striking parallel to those who believe we should just "let science speak"—that facts are facts and cannot be questioned—appears in the US law of evidence, specifically, in the idealization of science that accompanied the US Supreme Court's 1993 *Daubert* opinion, which introduced a new regime of judicial scrutiny with respect to the admissibility of expert scientific testimony in court.[19] The image of the scientific enterprise adopted in *Daubert* tended to both downplay the uncertainties in science and overemphasize the relation between scientific credentials and truth.[20] Those today on the "Because Science" side are likewise utterly confident in their facts and in the scientific establishment's delivery of those facts. That perspective overlooks not only the tentative nature of much of our scientific knowledge, its models, and its probabilities, but also the social character of scientific knowledge, that is, the cultural, linguistic, and economic supports of science. The latter do not necessarily reduce our trust in consensus science—it is the best we have—but they do offer us a basis for some humility when dealing with those one considers to be in the thralls of fringe science. In a strange turn of events, the idealization of science in law (in the context of courtroom expertise) brought on by *Daubert* has since waned due to the now-discredited identification techniques in forensic science; a more modest view of science, recognizing its limitations, has lately prevailed.[21] However, as I show in this book, that modest view of science has not yet become a dominant narrative in the crisis of expertise.

Returning to my four-part typology of experts (consensus scientists, their supporters, minority scientists, and their supporters), it is important to reemphasize that those ordinary, *citizen* "experts" who believe in consensus science, as well as those *citizen* "experts" who believe in minority-view science—it is part of their respective worldviews—have not, of course, become *scientists*; they are only experts in the sense of belonging to a community of believers with a shared conceptual language. Consequently, the respective worldviews of citizen-believers in consensus or in minority-view science is not the same as the respective worldviews of the *scientific* communities that produce those two different scientific positions. However, both producers are experts—the scientists on either side are experts—belonging to a community with a shared language—in some scientific field of research and discovery (whether consensus or minority-view science)—while the non-scientists are, respectively, "experts" in a system of beliefs by which they live their lives. Non-scientists trust their favored scientists, but they likely do not understand deeply that science, as they are not trained scientists. The non-scientist believers in a worldview (whether based on consensus science or otherwise) are like jurors in a trial with opposing scientific experts—the jury gets to decide which expert to believe, perhaps based on mere credentials, or perhaps on the basis of which expert was perceived as making a more compelling scientific presentation, but the jurors do not thereby become scientific experts.

The purpose of this book is to reinterpret the crisis of expertise (including the perception that half of the US population does not trust relatively consensus science) in terms of ideology or "quasi-religion." Importantly, I use the term "ideology" here not pejoratively, that is, not in some classical Marxian sense of a dominant ideology (perhaps one that would disappear in a classless society), but rather in the broader, descriptive sense of "a relatively well-systematized set of categories which provide a 'frame' for the belief, perception and conduct of a body of individuals."[22] Moreover, I use the term "religion" not in the sense of recognized world religions, or as a matter of worshiping a deity, but rather in the sense of a set of faith-like commitments that function as a worldview and an explanatory framework. (Therefore, I am not joining those who argue that the culture wars set Christianity against a "pagan" culture; in my view, Christians and secularists occupy both sides of the culture wars—liberal Christians are distinct from conservative Christians, and liberal secularists are distinct from conservative secularists.)[23] That we live in different worlds is, in my view, not simply a metaphor for contemporary disagreements, but an accurate description of the experience of different realities.

In Chapter 1, I acknowledge the complexity of the crisis of expertise, as well as the danger of assuming that only one "side" is operating ideologically. I then survey some recent explanations (of our crisis of expertise) offered by sociologists and endorse their suggestion that recourse to scientific explanations needs to be accompanied by some level of modesty. Finally, I address two institutions that some believe will solve the crisis of expertise, namely, the law and the scientific establishment. In other words, why not just impose scientific *truths* through legal regulation? Or, alternatively, why not just have a majority of scientists announce a compelling consensus that everyone should accept? I explain that both of these institutions have failed, and will continue to fail, due to (1) the current capture of the legal system by political parties and (2) the fact that the consensus of a majority of scientists alone is not necessarily compelling nowadays.

Chapter 2 focuses historically on the philosophical theory of quasi-religious worldviews in conflict by analyzing a 17th-century Dutch Golden Age painting and revisiting a unique 19th-century critique of Enlightenment claims to have risen above religious belief. Chapter 3 explores some recent studies that support my argument for an ideological or quasi-religious orientation to the crisis of expertise, including Professor Shi-Ling Hsu's identification of anti-science ideology in the Trump administration. Chapter 4 turns to religious belief as an example of expertise, drawing on: (1) the study of religious practices in the anthropology of religion; (2) the literature associated with the so-called Third Wave in science and technology studies (STS); and (3) Wittgenstein's concept of *forms of life*, including his remarks on the potential for communication across ideological or religious boundaries. I argue in Chapter 5 that viewing the crisis of expertise in terms of a battle between "expertises," grounded in faith-like commitments, will best serve the goals of communication, persuasion, and understanding in policy disputes in a divided nation. My analysis finds support in the literature associated with science communication theory and conflict resolution in policy disputes. However, I distinguish my approach from similar analyses by scholars in this field who, like Naomi Oreskes, emphasize a diversity of viewpoints as a key to trustworthy science and, like Frank Fischer, emphasize citizen science and deliberative collaboration in policy arguments. I conclude that while the culture wars generally seem intractable, there is hope for constructive and persuasive communication, based on understanding another's perspective and the values that drive their assessments of science, in the crisis of expertise.

My analysis goes beyond the familiar observations that a cultural distrust of expertise persists, that consensus science is worthy of acceptance, and that, in light of STS, we should be modest about consensus because science is tentative and often uncertain. Specifically, I adopt STS's Third Wave categories of expertise (explained in Chapter 4), extending that theoretical framework with a new constellation of disciplinary approaches—the anthropology of religion, the neo-Calvinist critique of the Enlightenment, and Wittgenstein's later philosophy (already part of the Third Wave)—to highlight the inevitability of ideological commitments in the crisis of expertise. Since conflicting expert communities have specialized languages and values that, respectively, unite and distinguish each community, the solution to the crisis of expertise involves persuasive communication between worldviews. That is made possible by: (1) self-awareness that we are all ideological; (2) modesty about the science in which one believes (that is, recognizing uncertainties); and (3) a common anchor in some level of scientific data and method (hence, my argument and solution do not apply to ordinary "culture war" controversies, such as who won the 2020 US presidential election). Before my analysis proceeds, however, it is important to: (1) confirm that distrust of expertise is sometimes justified; (2) distinguish "genuine" scientific controversies from those that do not really represent a disagreement between scientific points of view; and (3) identify several contemporary controversies, apart from the crisis of expertise, that are relevant to, and potentially benefit from, my analysis.

B. Justifiable distrust of (some) expertise

Sometimes, the call to distrust government experts comes from qualified industry or academic experts:

> High-risk Americans are having a tough time getting Covid-19 antiviral pills. ... The reason? Federal prescribing guidance is seen as vague. Some providers are reluctant to prescribe the pills due to uncertainty over how they interact with other drugs, how to apply the risk factors that qualify people for the treatment, and limited awareness on available supply.[24]

In the case of antiviral medicines like Pfizer Inc.'s Paxlovid or Merck & Co.'s molnupiravir, "some infectious disease professors say existing guidelines don't account for nuances in an individual patient's underlying conditions and medication profile."[25] Concerns with government expertise in this context would seem justified and therefore not exemplary of the crisis of expertise. Non-scientists, therefore, in some cases, "can have good reasons to question scientific expertise."[26]

Indeed, *pre-pandemic* studies of public perception of expertise identified "the presence or absence of expert consensus on the topic" as a factor informing judgments of expert trustworthiness—absence would seem to justify some level of distrust.[27] Likewise, distrust based on other conventional factors, including whether the expert (1) is perceived to be competent, (2) adheres to scientific standards, or (3) manifests integrity, appear justifiable.[28] In studies of public perception of expert trustworthiness *during* the COVID-19 pandemic, another highly influential factor was identified, namely, expert independence from political elites, which often interacted with other factors to shape public opinion—for example, lack of political independence was perceived as compromising integrity:[29] "In some cases, this interaction extended to people's assessment of [an expert's handling] of public communication; if an expert appeared in a media channel that the participants perceived as politically biased, or out of line with their own convictions, this negatively affected perceived trustworthiness."[30] In such studies, Scientists are therefore advised to "seek to maintain professional independence," to avoid politicizing expert knowledge, and to carefully consider "suitable channels of communication."[31]

Distrust of expertise based on the factors identified earlier is both justifiable and irrelevant to the crisis of expertise, which, in most cases, concerns distrust of consensus science. The important phenomenon called "pernicious epistemically justified distrust" of medical professionals—justified by immoral past actions of the professionals but currently harmful to the suspicious patient needing care[32]—is also not related to the crisis of expertise. This book is focused on: (1) the identifiable, growing phenomenon of *unjustified* distrust of mainstream scientific institutions and consensus science, whether on the basis of political affiliation, religious belief, feelings of estrangement, suspicions concerning elites, or even (with respect to climate change) fear that freedoms (including economic freedom) will be lost; (2) the communities that distrust consensus science; and

(3) how one might communicate with and persuade members of that community to change their minds.

C. On the question of *genuine* scientific controversies

Whenever someone argues, as I do, that we should adopt a modest view of science as tentative and often uncertain, and that we should therefore take seriously (and not reject out of hand) those who believe in minority scientific views, a frequent response is to question whether many of these minority views are "scientific" at all—in short, why should we be talking about ideology on both sides when one side is following science and the other is, in bad faith, "manufacturing" a controversy simply to call into question scientific consensus? As Ceccarelli states: "A scientific controversy is 'manufactured' in the public sphere when an arguer announces that there is an ongoing scientific debate in the technical sphere about a matter for which there is actually an overwhelming scientific consensus."[33] The idea of "manufactured" doubt was popularized by David Michaels in his book *Doubt Is Their Product: How Industry's Assault on Science Threatens Your Health*, which revealed the duplicity of tobacco companies.[34] Other common examples include AIDS dissent in South Africa's health policy debates, politically motivated global-warming skepticism, and the intelligent-design campaign against teaching evolution in secondary schools.[35] In the legal context, with respect to admissible expertise in the US courtroom, the controversy over "shaken baby syndrome" is another example—commentators identify an alleged shift in medical opinion, based solely on "outlier sources," that is used by defense attorneys to cast doubt on a "clinically valid and evidence-based [diagnosis], recognized by an overwhelming majority of" pediatricians.[36]

In Leah Ceccarelli's work in this field, she notes that rhetorically manufactured controversies often involve mercenary scientists who cherry-pick data to promote uncertainty.[37] Martin Weinel's similar analysis identifies four criteria by which we can differentiate between genuine and "counterfeit" scientific controversies:

1. Are the claims "conceptually continuous with science"?
2. Does the person making the claim have relevant expertise?
3. Is the claim supported "by some kind evidence"?
4. Is the claim subject to "an explicit argument" among scientists, or has it "reached the stage of implicit rejection" (that is, consensus)?[38]

If even one of the criteria is not met, the controversy is not genuine. Weinel concedes that the term "scientific controversy" is somewhat ambiguous, as it can refer to: (1) a controversy among scientists within their institutions; (2) a controversy in both science and the public realm (for example, how to end the COVID-19 pandemic); or (3) a controversy entirely in the public realm (that is, with no serious scientific disagreement).[39] Controversies in the first category are at least manageable by a commitment (on the part of disagreeing scientists) to norms of honesty; disagreements in the third category need not even be *called* scientific controversies. It is the second category with which this book is concerned. A problem identified by Weinel, however, is that in the second category, the debate can take place within science (with evidentiary standards to determine to whom one should listen) or outside science (where seemingly anyone can claim anything to anyone on any basis!).[40] Since my ultimate conclusion in this book rests, first, on leveling the playing field—we are all ideological—and then on persuading believers in a community of non-scientific "experts" (as noted earlier, expertise is not always correct) to change their beliefs, what, then, will be the evidentiary standard when the debate is not wholly within science? I am confident, however, that in matters concerning scientific expertise, citizens know how to recognize a scientist (whether a fringe or consensus scientist) and most have a concept that scientists amass evidence to make claims, often contrary to other scientists, about things like COVID-19 or global warming—that is, they recognize a controversy as scientific and take a side. The side they choose is a community of believers.

This book, therefore, is about arguably *genuine* scientific controversies, where there is at least fringe science and some data (even if minimal or weak) offered by honest scientists. For example, there is evidence that "scientists and experts (general practitioners, pediatricians, health care professionals and science journalists)" have seriously criticized "immunization policies and intervene[d] in the public debate" in Italy—this is not a manufactured controversy.[41] I am not discussing cases of fraud or bad faith in this book, and I would include in that category intentional "disinformation" (that is, not merely a mistaken belief in fringe science). Those who knowingly make fraudulent claims do not deserve the respect and understanding that my argument for persuasion presupposes. Likewise, my recommended communication between opposing worldviews is in terms of scientific inquiry—of data collected and interpreted—not in terms, for example, of astrology or magical powers.

One final example illustrates what I mean by "arguably *genuine* scientific controversies." The recent criminalization of medical treatment for transgender children in Texas and Alabama (on the basis that it constitutes child abuse) has been challenged as "biased science" because the medical claims justifying these new laws "are not grounded in reputable science and are full of errors of omission and inclusion."[42] The authors of that challenge identify two possibilities: the laws' drafters might genuinely misunderstand "medical protocols and scientific evidence," in which case, this may be a public scientific controversy (for example, an anti-trans fringe group is "repeatedly" cited); or their misstatements are deliberate and therefore fraudulent, indicating no genuine scientific controversy (the authors state that "[t]hese are not close calls or areas of reasonable disagreement").[43] This book is concerned with controversies in the first category.

D. Parallel controversies

I briefly identify in the following, without attempting to provide comprehensive analyses, three controversies in which the crisis of expertise appears and that are therefore related to the issues discussed in this book.

i. Conspiracy theories

Jaron Harambam, in his recent ethnographic study of conspiracy groups (in one of which he was embedded), notes that over 20 years ago, Peter Knight (in his own study of conspiracy culture) observed a loss of trust in science, a conventional epistemic authority: "In highly mediatized societies in which scientific disputes are played out in the open, it becomes increasingly difficult, Knight argues, to know 'which expert to trust—and how to decide if someone is indeed an expert.'"[44] Conspiracy theorists look for inconsistencies to destabilize pronouncements by scientific experts and often provide an alternative interpretation of the facts for their followers "to stage a contest over reality."[45] While conspiracy theories are not a focus in this book, Harambam's work led him to two conclusions relevant to my analysis. First, there are different types of conspiracy groups, so one cannot oversimplify and fail to see this difference between a collective belief that President Biden is a shapeshifting lizard, on the one hand, and a conspiracy theory about the government not revealing certain facts, which has happened, on the other. In the case of the latter type of conspiracy theory, we should therefore be more modest about what we know and less arrogant if we disagree:[46] "It is ... simply untenable," Harambam warns, "to argue that the belief in conspiracy theories is by definition delusional and paranoid."[47] Second, we need to explore the *meaning* of a conspiracy for its believers, by which Harambam means the cultural threats to their values *and their worldview* that potentially drive their beliefs. For example, former President Trump's 2016 "election theme—political, economic and cultural elites willfully work together to set aside the interests of ordinary Americans in favor of their own establishment benefits"—is not only a "conspiracy theory *par excellence*," but also an indication that many citizens feel alienated, making it easy to believe that the conspiracy is real.[48] My recommendation in this book that we need to understand those who believe in fringe science includes a need to see the *meaning* of consensus science—in controversies over vaccination or global warming—for those who might fear loss of freedom or economic loss from that science.

ii. Political-economy impacts

Sometimes, political goals can get in the way of scientific expertise—think of neoliberal arguments for small government and less regulation to allow the free market to solve social problems: "There is also a politics of expertise. Fossil fuel interests may buy their own researchers or research institutes … in order to avoid regulation. That is a clear subversion of science, an intrusion of politics in the raiment of objectivity."[49] In the previous Trump administration in the US, concerns arose that consensus scientists were removed from regulatory agencies in an effort to relieve businesses from excessive environmental regulation. Newly appointed fringe scientific advisors doubted the seriousness of global warming, The discretion given to agencies to decide how aggressively to enforce laws was thus a tool in the hands of corporate interests. This is a clear example of politics—the political perspective of one party in power— eclipsing consensus science.

However, in the critique of the Trump administration as anti-scientific, an idealized narrative of science appeared: Trump was ignoring scientific *certainties*. That narrative exacerbates the divisions of the crisis of expertise—critics of Trump should have acknowledged the uncertainties in their own risk analysis *and* appealed to their current models, probabilities, and data, as the fringe scientists were at least claiming a scientific basis for their views.

iii. Data as meta-expertise

A third context in which the crisis of expertise appears is the growth of meta-expertise: "the development of standardized and specialized algorithmic and quantitative methods that ... evaluate outcomes in particular experts' fields."[50] Based, in part, on distrust of experts and their disagreements, meta-expertise involves outsourcing a controversy for evaluation to consultants, economists, and quantitative analysts who promise to measure or rate the value of each expert and attain higher or more rigorous knowledge of the controversy: "[T]he crises of expertise ... have empowered meta-experts as potential agents of adjudication and resolution when mere experts' views are contested."[51] Implicit in this movement is also some level of distrust of human judgment, the response to which is often that decision making cannot be reduced to measurables; there is, however, a justifiable belief that algorithmic rules can, in some cases, correct human errors.[52]

Frank Pasquale has identified some of the limitations on meta-expertise, including the ability to game rating systems (the "choice of one set of data points for quantitative judgment excludes others"), the danger of shutting down "political debate or judgment," and the fact that meta-expertise "will always be dependent on the observation and dedication of persons charged with collecting and interpreting the data on which it depends."[53] Moreover:

> Scholars have spent decades working to draw attention to the normative commitments expressed in and through technology. ... Within computer science, the machine learning and mechanism design communities have been particularly active in ... heeding the call for attention to values, politics, and "social good" more generally, holding more than twenty technical workshops and conferences ... on some variation of fairness, bias, discrimination, accountability, and transparency in the last five years [since 2020].[54]

However, for "true believers in expansive meta-expertise, problems with existing metrics of assessment and algorithms of replication are simply a technical issue to be fixed, rather than a political problem to be negotiated and discussed."[55] A good example of such corrective efforts in the field of criminal law is the recent proposal to critically evaluate software tools used nowadays to convict an accused, even though defense counsel cannot cross-examine the "evidence":

> This offends the commitments of the adversarial criminal legal system, which relies on the defense's ability to probe and test the prosecution's case to seek truth and safeguard individual rights. ... [W]e propose robust adversarial testing as a framework for questioning evidence output from statistical software, which range from probabilistic genotyping tools to environment audio detection and toolmark analysis.[56]

Concerns remain with defense access to data and expert witnesses to accomplish this, but at least it shows a response to meta-expertise that recognizes its value but attends to its limitations.

As a "solution" to the crisis of expertise, the movement toward meta-expertise may do little to remedy the cultural divisions that result in distrust of consensus science—the subject of Chapter 1. The pronouncements on the basis of meta-expertise may face the same fate (as consensus scientific experts) if they are viewed by large segments of the population as the voice of alleged elites, arrogant technocrats, or a government for which those citizens did not vote.

ONE

What Caused, and How Do We Fix, Our Crisis?

A. The complexity of the crisis of expertise

In the culture wars generally, there is a tendency to oversimplify; for example, some critics often talk of those who live in an "alternate reality" as if they—the critics— live in the "real" world. As to some issues, this may be justified. Stories of those who believe their parents have been brainwashed by Facebook posts claiming election fraud (or a grand left-wing conspiracy), or who cannot believe their friends supported an "authoritarian" president who divided the country, are common.[1] However, many of the differences between the two sides in the culture wars reflect different values and visions for the nation. As sociologist Nissim Mizrachi explains, with respect to the working-class voters in Israel who supported Netanyahu (the parallels with Trump voters "are impossible to miss"): "The problem [is not that they were] confused about what was best for them. They weren't suffering from … 'false consciousness'. … [They] were consciously spurning liberalism for a reason: what they see as the endgame of the liberal worldview is not a world they wish to inhabit."[2]

One need only think of the abortion rights controversy, immigration policies, or the supposed attacks by the Left on religious freedom or gun rights to recognize that the division in the culture wars is not simply about misinformation from unreliable Internet sources; it is about values, identities, and foundational commitments to a way of life. Moreover, those on the Right do not have a monopoly on foundational commitments—both the Left and the Right have a moralized anchor "around which to understand the world":[3]

> The delusional claim not to have any ideology ... is almost always a camouflage. Just as in the joke about one fish saying to the other "what's water?" ... the claim not to have any conscious ideological positions at all signifies at best that [one] has simply absorbed the dominant ideology.[4]

One can certainly argue that there is no moral equivalency between the opposing ideologies in the culture wars and that the views of one side will lead to superior outcomes in terms of fairness in opportunities for success, racial and gender equality, or helping those in need. However, the notion that any group occupies a neutral center in politics is a myth.

Likewise, with respect to scientific facts in the crisis of expertise, some humility is also warranted. The "Because Science" and "Trust the Science" T-shirts worn by those on the Left who see the culture wars as a struggle between those with apolitical scientific facts and anti–scientist ideologues are also oversimplifications. Even with something as clear as the much-needed COVID-19 vaccine, there are:

> all kinds of extra-scientific variables: moral assumptions about what kind of vaccine testing we should pursue (one reason we didn't get the "challenge trials" that might have delivered a vaccine much earlier); legal assumptions about who should be allowed to experiment with unproven treatments; political assumptions about how much bureaucratic hoop-jumping it takes to persuade Americans that a vaccine is safe.[5]

It is never as simple as "following the science." If one assumes science is always right, that position is easily attacked by highlighting the publication by *The Scientist* of the "Top retractions of 2020":[6]

> The cliché is that people should "follow the science" and do whatever "science says." But the truth is that science says many things at once. Science says that the coronavirus can last one month on surfaces; it also says it's vanishingly rare to get the coronavirus from surfaces. Bad studies, good studies, and mediocre studies are all part of the cacophonous hydra of "science" that is constantly "saying" stuff.[7]

To build trust in science, one needs to be honest about its limitations.

B. (Dis)Trust in science

Bruno Latour's *Down to Earth: Politics in the New Climatic Regime* was prompted not only by the election of President Trump in 2016, but also by the growth of three phenomena in advance of the election, namely: (1) deregulation, (2) human inequalities, and (3) "a systematic effort to deny the existence of climate change."[8] The latter phenomenon has resulted in the loss of a shared, *common* world[9]—consensus science revealed the new climatic regime, but that science is the target of climate-change deniers.[10] The "deniers" include the people who are suspicious of elites, fearful, feeling betrayed, living "within a fog of disinformation," and therefore supporters of Trump. Importantly, "No attested knowledge can stand on its own, as we know very well. Facts remain robust only when they are supported by a common culture, by institutions that can be trusted, by a more or less decent public life, by more or less reliable media."[11] However, Latour observes, our "rational" journalists not only identify consensus science "deniers" as being naively attached to "alternative facts," but also "continue to believe that facts stand up all by themselves, without a shared world, without institutions, without a public life, and that it would suffice to put the ignorant folk back in an old-style classroom … for reason to triumph at last."[12] Journalists do not seem to realize that *they* "live in an alternative world," where there is climate change—"there are now several worlds … and they are mutually incompatible."[13]

That theme of different worlds is echoed in mainstream media and among scholars focused on the crisis of expertise. For example, Jan Lytvynenko, a reporter for *BuzzFeed News* who is based in Toronto, identified (on CNN) what he referred to as Trump's "numerous falsehoods" as being "part of an ecosystem environment. They're part of hyperpartisan news websites, of commentators and influencers who support the president who also are very actively engaged in spreading false information that social media networks might not always catch."[14] Likewise, on the final episode of season 18 of HBO's "Real Time with Bill Maher," guests Alex Wagner (journalist and co-host of "The Circus" on Showtime) and presidential historian Jon Meacham discussed with Maher their sense that Trump supporters live in a parallel universe with "alternative" facts; partisanship was referred to as "religious," and the right wing as a "cult."[15] Finally, in a *New York Times* story about family rifts in the Trump years, one woman who called her parents "brainwashed" was told by her father that *she* was brainwashed, while another who argued with her mother remarked, "This is not even a political divide, it's a reality divide."[16] It should be noted that popular television and print news stories often include references not only to different worlds, but also to the corresponding religious zeal of those caught up in an alternative reality.

With respect to *scholarly* responses to the crisis in expertise, a Mellon Sawyer seminar addressing "Trust and Mistrust in Experts and Science during the Pandemic" was convened by a team of scholars, including Columbia sociology professor Gil

Eyal. The October 29, 2020, session, entitled "Experts, Publics and Trust During the Pandemic: Sociological Perspectives," included presentations by University of California—Los Angeles sociologist Rogers Brubaker, Cornell sociologist of science Steve Hilgartner, University of Southern California sociocultural anthropologist Andrew Lakoff, and University of North Carolina sociologist Zeynep Tufekci.[17] Eyal, in *The Crisis of Expertise*, highlighted how, in an age so dependent on science and expertise, we are seemingly witnessing "increased suspicion, skepticism, and dismissal of scientific findings, expert opinion, or even whole branches of investigation."[18] One problem for scientific expertise in regulatory contexts, Eyal notes, is that scientists cannot simply appeal to "facts"; rather, they have "estimates, models, predictions, forecasts, points on a graph, [and] expert judgments."[19] The latter, the judgments made in contentious policy disputes, are "trans-scientific" ("Expertise is our name for this realm"); overcoming the inevitable incompleteness of data requires "presuppositions, assumptions, 'priors,' heuristic conventions, 'acceptable levels,' cutoffs, and so on."[20] In the end, to rebuild confidence in science, Eyal seeks an institutional solution: a republic of trans-science, *not* based on the model of conventional science or on political think tanks (or even on the democratization of science through lay "expertise"), but instead an institution similar to US Supreme Court Justice Breyer's notion of an elite civil service corps of interagency experts.[21]

That ideal of neutral experts has a history in the law of expert evidence, especially as a solution to the persuasive charlatans or advocacy-oriented experts in court—the same problem that the US Supreme Court's 1993 *Daubert* opinion was supposed to solve.[22] While I think the ideal of neutral expertise is a misleading idealization of science (that is, it presumes there is such a thing as a "neutral" science)[23]—unwitting in Eyal's case, as he is obviously well versed in the social and value-laden aspects of science—Eyal is certainly correct that the institution of science itself needs new support.[24] Latour includes *institutions*, alongside culture and the media, in his list of social structures that support "attested knowledge."[25] Collins and Evans, in *Why Democracies Need Science*, also emphasize the importance of trusted social institutions: science is a potential example, and as a *social* institution, it has the capacity for moral leadership.[26] In a pluralistic democracy, scientific expertise can be one of the checks and balances when a presidential administration ignores scientific evidence.[27]

At the aforementioned Mellon Seminar, each panelist addressed, from a sociological perspective, not only the frayed relations between the public and experts—the lack of trust in scientists—but also the disagreements between experts themselves that enable the politicization of science and amplify divisions in society.[28] Rogers Brubaker emphasized the usual challenges of populism with respect to expertise, including populism's anti-institutional and anti-intellectual aspects, which paradoxically (because expertise was *needed*) grew more intense during the COVID-19 pandemic.[29] Common sense (or lay expertise) is "valorized," complexity is distrusted, announced crises seem overblown as things get better, and the accessibility of a great deal of data makes it easy to find inconsistencies among expert reports.[30] The notion that "experts don't agree on anything" leads to both the "democratization of the means for assessing expertise" and a multiplication of pseudo-experts.[31] The digital era exacerbates this democratization, undermines epistemic authority, and rewards populist styles of confrontational and simplistic communication[32]—"anyone with a web browser and an internet connection [can] publish."[33] Brubaker concludes, in terms most relevant to this book, that we inhabit "radically different public worlds," constituted by what each side knows and believes, with no shared definitions of, for example, the COVID-19 pandemic.[34] The institution of science thus appears fragile, not robust, nowadays.

The next panelist in the Mellon Seminar, Steve Hilgartner, highlighted the "idealization" (my term) that is evident in the half of the "public world" that reveres consensus science.

Hilgartner's example is the popular account of the (unopened) pandemic playbook, based on scientific expertise, that President Obama (and President Bush) left in the White House to ensure that the US would be prepared for the next—our present—pandemic.[35] That playbook, and its science, were not followed, the story goes, hence the incompetence which caused unnecessary deaths in the US in 2020. The playbook, Hilgartner explains, imagined a world with a government attentive to expertise, willing to provide resources to activate such a plan, and in communication with citizens willing to listen and obey experts.[36] That ideal image, however, ignores not only the difficulties of a divided culture where knowledge and politics are inseparable, but also the way governments and officials have to deal with new "facts" and data—there are multiple interpretations, degrees of risk and uncertainty, and conflicting values in policy and in science itself.[37] In effect, the optimistic, reassuring narrative (that it was only the failure to follow the plan that caused our problems) excludes a counter-narrative that disasters are inevitable and not always manageable, that playbooks have limits and cannot predict every problem, and that there is no mythical managerial world:[38]

> In this context, claims about a lack of trust in expertise have to be understood not only as descriptive statements [of] the world in which we live, but also as performative ones, statements intended to act on the world … in a variety of ways, by discrediting adversaries, mobilizing support, reasserting the need for expert authority, and … reassuring people that technocratic modes of governance can be made to work if only the citizens and the leaders would fall in line behind a reliable and apolitical science.[39]

I highlight Hilgartner's comments not to suggest that scientific expertise should not be trusted, but rather to argue that those who claim that half the country is ignoring expertise are also part of a community of believers—they believe strongly, even idealistically, in the accuracy of their narrative and their "world." Notably, that *other* half who ignore consensus experts have their own beliefs concerning why experts should *not* be trusted, including concerns that experts are too dogmatic, overly authoritative, unable to reflect critically on their own "practices and roles," and even dangerous in a democracy with "deliberative ideals."[40]

Most of the panelists at the Mellon Seminar emphasized the need for some modesty on the part of those who live in the world of "cold, hard facts" and "Because Science" T-shirts. First, there is concern that scientists are somewhat arrogant, such that the public's mistrust of science is matched perfectly with scientists' distrust of the public—and this is not helping the crisis of expertise:

> Public health officials have sent confusing messages about Covid policy. They have done so on masks, tests, adult vaccines and basic Covid statistics. Sometimes, the confusion has been intentional [and] has become one more factor contributing to Americans' distrust of major institutions like the government, the media and the medical system. ... Public health officials in this country are often uncomfortable trying to convey the full truth. They ... provide only partial truths and hope that Americans won't notice. The strategy hasn't been very successful.[41]

Zeynep Tufekci, in her Mellon Seminar remarks, argued that one of the reasons that various governments' pandemic response has been challenging is due, for example, to the common-sense fears on the part of the public that: (1) masks provide a false sense of security, which might cause people not to be careful; and (2) masks might, early in the COVID-19 pandemic, be hoarded if experts continue to overemphasize mask wearing—both of which are at least understandable.[42]

The response to these fears on the part of scientists is that citizens cannot be trusted to wear masks, and the public feels that mistrust; we therefore have a mutually reinforcing cycle of distrust, exacerbated by social media in our digital age.[43] One conclusion is that in public health matters, one cannot simply insist that everyone listen to the technocrats—experts must defend their scientific claims, while acknowledging both their occasional failures and the inevitable uncertainties in the scientific enterprise.[44]

Moreover, as Andrew Lakoff mentioned in response to a question at the Mellon Seminar, there is evidence from the sociology of disasters that the public should be listened to in some contexts.[45] The final conclusion of the Mellon Seminar, perhaps offering a partial solution to the crisis of expertise, was to make the contingencies of science, that is, its failures and uncertainties, more visible to the public. Steve Hilgartner said citizens should know that science is a process and not a producer of unchangeable facts.[46] Rogers Brubaker added that expertise is also complex because even if there were such a thing as perfect laboratory science, once experts enter the public realm, they are not able to answer all important questions; therefore, we need to "reduce public expectations" of expertise, publicize the "intrinsic limitations" of science, and adopt a more modest approach.[47]

The Mellon Seminar highlights my argument that we should not view the crisis of expertise as a conflict between those who operate based on faith in social media conspiracies and myths, on the one hand, and those who possess the truth that is delivered to them by the scientific establishment, on the other. Rather, the conflict is between believers—between those who put their (respective) faith in different experts, political parties, media sources, and narratives about who can be trusted. It is often acknowledged, in references to right-wing citizens, that "partisanship today doesn't just mean excessive loyalty to a party and its program," but also "implies a kind of secular faith."[48] Based on the Mellon Seminar's concluding proposal that those wearing "Because Science" or "Science is TRUTH" T-shirts might adopt a more modest approach, I suggest that those T-shirts likewise imply, on the side of those who trust consensus science, "a kind of secular faith," such that the crisis of expertise is a quasi-religious, and not only a scientific, controversy.

C. The failure(s) of law

Readers may wonder whether the solution to the crisis of expertise lies in better legal regulations, based on sound science. My initial response is that legal initiatives are sometimes ineffective:

> The evidence suggests that broad mask mandates have not done much to reduce Covid caseloads over the past two years. Today, mask rules may do even less than in the past, given the contagiousness of current versions of the virus. And successful public health campaigns rarely involve a divisive fight over a measure unlikely to make a big difference.[49]

My next response is to point out that the law is subject to politics. While there may have been a conventional commitment to consensus science in governmental agencies, career advisory scientists can be fired and replaced by scientists with fringe views on such matters as global warming. Legislation requiring that consensus science be followed would always be subject to judicial interpretation or invalidation (and to arguments that there is no consensus) and to repeal by a newly elected legislature. If conservative politicians want "to strip the Environmental Protection Agency of its authority to regulate planet-warming gas emissions from power plants," a conservative Supreme Court might agree.[50] The US Centers for Disease Control's (CDC's) masking requirements on mass transit might be successfully challenged in court, weakening that agency's authority:

Judge Kathryn Kimball Mizelle of the US District Court for the Middle District of Florida ruled that the CDC incorrectly described the mask mandate as a form of "sanitation" to justify its authority in the matter. She questioned why the CDC didn't look for alternatives and said the order doesn't actually require universal masking to stop transmission since it allows for exceptions, such as people who are eating or drinking.[51]

I therefore have doubts about a legal solution to the crisis of expertise.

Many commentators have highlighted *Jacobson v Massachusetts*, where the US Supreme Court rejected a challenge to smallpox vaccine requirements, as an example of law following science.[52] The US Supreme Court held:

> [T]he liberty secured by the Constitution of the United States to every person within its jurisdiction does not import an absolute right in each person to be, at all times and in all circumstances, wholly freed from restraint. There are manifold restraints to which every person is necessarily subject for the common good. On any other basis, organized society could not exist with safety to its members.[53]

One can object that this opinion reflected and should be limited to the extreme terror of the smallpox epidemic, but a greater challenge is Professor Scott Burris's concern about the "viability" of *Jacobson* now that recent "courts have unveiled a new view based less on the social contract than on a strong form of libertarianism."[54] Some courts during the pandemic have indeed followed *Jacobson*, as in the challenge to Indiana University's COVID-19 vaccine mandate,[55] but other states with conservative governors prohibit mandates by private employers—revealing "an odd libertarian streak that dislikes government orders to individuals but then says private employers shouldn't be free to choose."[56] For example, on May 3, 2021, the Florida governor signed into law a prohibition against businesses requiring vaccine documentation for entry.[57] Even that law can be challenged—on August 9, 2021, Norwegian Cruise Line's parent company, pursuing a policy of requiring passengers to be vaccinated on its three cruise lines, was granted a preliminary injunction against Florida's prohibition on vaccine mandates, but such victories can be overruled by a higher court.[58] Those opposed to vaccine mandates argue that enforcement is problematic due, for example, to fake documentation or to difficulty in evaluating requests for religious (or philosophical, in some states[59]) or medical exemptions. However, the major factor destabilizing attempts to require vaccination in retail establishments and places of employment is that the courts cannot be relied upon consistently to settle controversies between consensus and fringe scientific beliefs.

Sometimes, the legal system seems to support consensus science, as in the recent case of parents suing a school system for violating their children's constitutional rights by imposing a mask mandate—not only was the suit dismissed for lack of a credible legal claim, but the judge ordered the parents to pay $57,000 in attorneys' fees to the school system for its defense costs.[60] On the other hand, when the US Biden administration recently tried to mandate COVID-19 vaccines for Air Force and Air National Guard servicemembers, a federal judge blocked the mandate because requests for religious exemptions were "sweepingly rejected"—and, of course, the "government is expected to swiftly appeal this decision."[61] Support for consensus science in legal contexts is clearly not uniform.

A related legal effort in the context of litigation involving scientific issues is the proposal for a new evidentiary rule that would condition admissibility of expert testimony in the courtroom on evidence that the testimony is grounded in consensus science.[62] The proposal has the distinct advantage of taking the decision as to which side's expert to believe away from judges and juries not trained in science, and shifts the question to one a bit more comprehensible to judges and juries—which expert is in sync with the majority of scientists in their field? However, the proposal assumes that juries in our divided culture will respect and follow consensus science. My analysis of the crisis of expertise demonstrates: (1) that consensus is not easily discerned by the public; and, more importantly, (2) that one cannot assume respect for consensus, which must be earned by persuasion and mutual respect. Both sides in public scientific controversies "announce" that their respective views are correct. For that reason, as I discuss in the next section, consensus scientists alone cannot be relied upon to solve the crisis of expertise.

D. The failure(s) of science

One *encouraging* feature of the crisis of expertise is the apparent trust, on both sides of the controversy, in scientific research and researchers:

> [P]ublic debates rarely feature open resistance to science; the parties to such disputes are much more likely to advance diametrically opposed claims about what the scientific evidence really shows. The problem, it seems, is not that members of the public are unexposed or indifferent to what scientists say, but rather that they disagree about what scientists are telling them.[63]

Research into the cultural cognition of *risk* indicates that individuals "credit or dismiss evidence of risk in patterns that fit values they share with others," with the research extended "to evidence of what scientific expert opinion is on climate change and other risks."[64] We therefore cannot assume that individuals with diverse outlooks will agree on what scientific consensus is: "[C]ultural cognition influences perceptions of credibility. Individuals more readily impute expert knowledge and trustworthiness to information sources whom they perceive as sharing their worldviews and deny the same to those whose worldviews they perceive as different from theirs."[65] Individuals also "tend to search out information congenial to their cultural predispositions," and to "systematically overestimate the degree of scientific support for positions they are culturally predisposed to take."[66] Therefore, scientific consensus established by agreement among experts is not likely "to counteract the polarizing effects of cultural cognition"—those with diverse worldviews will not assess the state of consensus in the same way.[67] Importantly, these conclusions are not a description of one side in the culture wars and crisis of expertise, but involve both sides—everyone has a worldview, which "tends to generate conflict in public deliberations."[68]

The authors of the foregoing study are optimistic about the possibility of communication ("rational public deliberations") in the context of diverse worldviews if we recognize that the "enfeebled power of scientific opinion" is *not* due to failure of experts to agree and disseminate their opinion:

> [C]ommunicators must attend to the cultural meaning as well as the scientific content of information. ... When shown risk information (e.g., global temperatures are increasing) that they associate with a conclusion threatening to their cultural values (commerce must be constrained), individuals tend to react dismissively toward that information; however, when shown that the information in fact supports or is consistent with a conclusion that affirms their cultural values (society should rely more on nuclear power), such individuals are more likely to consider the information open-mindedly.[69]

Likewise, individuals "reflexively reject" experts with values opposed to their own but become open-minded "if they perceive that there are experts of diverse values on both sides of the debate."[70]

This analysis parallels the concerns in Frank Fischer's *Truth and Post-truth in Public Policy*:[71] "It is not that facts are unimportant. Rather, it is that they gain meaning in the policy world from the social and political contexts to which they are applied. Thus, the social-subjective meanings that factual information have for political participants need to be brought into the analysis."[72] Contrary to the views of some climate-change scientists, better facts may not initially be effective in changing the minds of those worried about loss of freedom and economic distress: "meanings drawn from ideological value orientations are used discursively to interpret factual data in denial arguments."[73]

These insights are significant for my proposal that communication across worldviews concerning scientific matters is possible, but only if one: (1) concedes that we all have worldviews; (2) acknowledges with modesty the tentative nature of any scientific consensus; and (3) respects any appeal to scientific data and argument. Those who only emphasize, in their diagnosis of the crisis of expertise, the need for scientists to clearly communicate scientific consensus misunderstand the importance of cultural cognition. Likewise, those who recommend going one level higher, that is, to a non-ideological committee of "neutral" scientists to settle scientific controversies, have unwittingly idealized the scientific enterprise as if it is possible for one group of scientists to rise above the theoretical, political, social, cultural, linguistic, economic, and institutional aspects of science. Scientists on both sides are political, and to assume one is not, as Gilroy-Ware warns, is risky: "When your arrogance ... means that you fail to critique your own political beliefs, you leave yourself open to others formulating that criticism for you, usually with a lot more antagonism."[74]

The aforementioned research into cultural cognition is particularly helpful in this regard, except for the authors' thin, imprecise concept of "worldview," almost as if it needs no careful explication. Therefore, the starting point for my diagnosis of the crisis of expertise is the recognition of the almost "religious" nature of worldviews that affect the reception of scientific information by the public.

TWO

Worldviews
as "Religious" Frameworks

Over two centuries ago, Thomas Paine confirmed that "religion" need not be associated with deistic beliefs and rituals: "The word religion is a word of forced application when used with respect to the worship of God. The root of the word is the Latin verb ligo, to tie or bind. From ligo, comes religo, to tie or bind over again, to make more fast."[1] In an effort to clarify a notion of "religion" as being analogous to any set of beliefs, values, and commitments that govern one's life—what I am calling a "worldview" or "ideology"— it helps to revisit and consider the culture wars in Holland during the Protestant Reformation (16th–17th centuries) and continuing through the 19th century. There is a painting in the Metropolitan Museum of Art that reminds us of the two different worlds that Christian believers—Protestant and Catholic, respectively—occupied in 17th-century Holland.[2] Golden Age art historians have noticed, both in that painting and in others, the Calvinist perspective on religion as a worldview affecting all of one's life, rather than having to do primarily with worship, ritual, or prayer. In the 19th century, neo-Calvinist politicians used that unique conception of religion to argue that Enlightenment rationalism, as well as the ideology behind the French Revolution, were both religious in nature, involving beliefs that paralleled (and opposed) the Calvinists' religious beliefs.

A. Revisiting a Dutch Golden Age painting

Wholly apart from complex doctrinal disagreements and the hidden prayers of believers, the religious fervor of the Reformation had a visual aspect:

> The whitewashed walls of … Calvinist churches vividly call up the historical re-formation of religious space. … This type of space has been purified; as past visual practices were redefined as idolatry or superstition, it has been emptied of images, circumscribed by Calvinist prohibitions against the … reception, or veneration, of imagery.[3]

My initial focus in this chapter is on the representation of two important features of Dutch Calvinism in the painting "Interior of the Oude Kerk, Delft," a detailed study by Emanuel de Witte (1616–92) of an 11th-century, formerly Catholic church, with whitewashed walls and no images of Christ, nor of Mary or any other saint.[4] First, the Calvinist idea that *all* people (not only the clergy) are called by God to hold an "office" suggests that all of life is religious, even business or farming (and not only Sunday worship). Second, the Calvinist notion that all individuals have access to the scriptures and therefore to God—without mediation by clergy[5]—likewise takes religion outside the church and into the everyday life of believers. Those two features—indicating that the Calvinists lived in a different world than Dutch Catholics—are suggested in de Witte's painting by: (1) the iconoclastic cleansing of the church (of Catholic imagery) that preceded the painting; (2) the omission of the pulpit and the civic banners that decorate the space, which seemingly degrade and transform the sanctity of the church into a different kind of meeting place; (3) two children scribbling on a column, and two dogs, one urinating on another column; (4) two merchants who appear to be transacting business; and (5) a man talking to a woman and child, perhaps a husband and father, or just a friend. This human (and canine) scene could be in a park or town market, but these figures in the church imply neither (1) disrespect of Christianity, nor (2) secularism overtaking a religious space. Quite the contrary: the disrespect is reserved for the papacy. Far from any triumph of secularism, Dutch Calvinism is an argument for the religious character of *everything*.[6] Merchants doing business, children playing, even a dog urinating are not relegated to an arena that is secondary to some holy space. There is no division between a spiritual realm and a natural world; there is just a world in which believers live.[7] While there is no doubt that Catholics are believers as well, they are indirectly treated as simply wrong about what Christianity entails. This is the Calvinists' world and their unique worldview—and, of course, their understandings of their God and their religion is not the worldview of their Catholic neighbors.

There is therefore more going on in this painting than the skill of a renowned architectural painter, a genre painting of everyday life, or a picture of the actual church; the painting depicts the results of Calvinism as a collective ideology.[8] It is, in that respect, a historical document, notwithstanding its inevitable fictive character (it is a typical, but not an actual, scene), concerning the effects of the Reformation in the north of Holland in the 17th century and thereafter. In all likelihood, the artist is *not* intentionally instructing us on Calvinist notions of individualism, or promoting the view that all of life is religious.[9] Rather, the images unwittingly reveal a set of meanings that are implicit in Dutch Calvinism,[10] which is, by its own admission, an ideology.

In my interpretation of de Witte's painting of the Old Church in Delft. I am particularly interested in John Calvin's (1509–64) condemnation of the images historically associated with Christianity and present (only temporarily) in Roman Catholic churches and cathedrals (including St. Peter's, the cathedral Calvin used in Geneva). Calvin "found all image veneration misguided, as God's divine power could not be harnessed through visual representations,"[11] hence the purging "of icons and religious imagery," as well as the hiring of "painters to cover the wall and vault paintings in order to accommodate the new worship practices of the Reformed congregations"—a century before de Witte's 1650 painting of the Old Church.

The *meaning* of this effort to take over and cleanse the Catholic churches in northern Holland is not simply a theological rejection of Catholic imagery, but a new conception of religion. Some Dutch Calvinists, for example, emphasize John Calvin's legal training and political acumen—one need not be in the clergy to be in a spiritual profession.[12] For Calvin, all aspects of life, not just those conventionally "religious" matters like church attendance or prayer, are equally and significantly "spiritual."

B. From a painting to a theory of "religious" worldviews

Perhaps surprisingly, there is a direct line between Golden Age church interiors and the French Revolution:

> In 1789, the turning point was reached: "We no more need a God" ... heralded the liberation ... from all Divine Authority ... There is no doubt then that Christianity is imperiled by ... serious dangers. Two life systems [modernism and Christianity] are wrestling with one another ... This is the ... struggle for principles in which [Holland] is engaged.[13]

The very same ideas that are depicted in de Witte's painting continued to influence Dutch Calvinism. The Calvinist notion that each believer's faith influences and directs everything they do became the basis for the notion that *un*believers (or Catholic *mistaken* believers) must also have a worldview, an ideology with a religious (that is, belief-based) character, which influences their respective public and private lives. The contrary notion that human beings live on the basis of reason, whether based in

Greek philosophy (especially Aristotelean), Catholic doctrines of faith and reason (especially Thomistic),[14] or Enlightenment rationality, is rejected as a failure to see the inevitability of belief structures.

Neo-Calvinist Guillaume Groen van Prinsterer (1801–76), eventually the leader of Holland's Anti-Revolutionary Party, was a critic of the Enlightenment ideas that led to the French Revolution (he called it a "Reformation in reverse").[15] Groen's argument that a "religion of unbelief" was at war with Christianity leads to an ideological conception of religion: it is not belief in, or worship of, a divinity that makes a religion, but a framework of foundational beliefs that guide the lives of believers.[16] "Religion" is therefore more like an ideology.

Abraham Kuyper (1837–1920), Prime Minister of Holland (1901–05) and a Dutch Reformed Church pastor, was Groen's successor both in parliament and as leader of the Anti-Revolutionary Party. In his 1898 Stone Lectures at Princeton, Kuyper described Calvinism as a *Weltanschauung*—a religion for all of life (alongside the competing "religion" of modernism)—affecting one's perspective on all matters.[17] Calvinism embraces not only theology and worship, but also politics, science, and art. Inheriting Calvin's emphasis on individual rights (for example, freedom of association and liberty of conscience), Kuyper also developed a political theory of "sphere sovereignty," whereby under God's sovereignty, church and state were sovereign within each's sphere of competence and neither had authority over the other.[18]

Kuyper's disciple, Herman Dooyeweerd (1894–1977), trained in law and later the chair in jurisprudence at the Vrije Universiteit Amsterdam (founded by Kuyper), expanded the notion of *two* conflicting belief systems (which he called *grondmotieven*) to *four*, roughly, Greek, Catholic, Enlightenment, and Biblical. The first three, as explained in Dooyeweerd's magnum opus *De wijsbegeerte der wetsidee* (translated as *A New Critique of Theoretical Thought*), share a commitment to the autonomy of human reason (respectively, for example, developed by Aristotle, Aquinas, and Descartes), seemingly rational and therefore neutral.[19] Dooyeweerd, on the other hand, in a *transcendental* critique that echoed neo-Kantianism, discerned pre-theoretical, conscious or unconscious, ideological commitments on the part of all of these *believers*. As to the Biblical worldview, Dooyeweerd confirmed his Dutch Calvinist heritage by arguing that "a radical Christian philosophy can only develop in the line of Calvin's religious starting-point."[20] Dooyeweerd therefore concedes his own ideological commitments, but he does so within a philosophical tradition in which we are all, inevitably, believers: "I do not pretend that my transcendental investigations should be unprejudiced. On the contrary, I have demonstrated that an unprejudiced theory is excluded by the true nature of theoretic thought itself."[21]

This rejection of the rational, Enlightenment subject sounds postmodern, and there are parallels between Dooyeweerd's critique of ideology and the contemporary identification of social influences on, even social construction of, the natural sciences. The natural sciences do indeed provide stable knowledge, but not because they escape or rise above ideology. For Dooyeweerd, all the "sciences"[22] reflect pre-theoretical commitments or belief structures like those variously identified by many scholars in 20th-century history, philosophy, and sociology of the natural sciences. Scientists should certainly avoid personal religious interference with their research, but they cannot avoid the theoretical, social, linguistic, and economic structures that make science possible: "That [any] critical investigation is necessarily dependent upon a [supra-] theoretic starting point does not derogate from its inner scientific nature. The latter would only be true if the thinker should eliminate a ... scientific problem by a dogmatic authoritative dictum, dictated by his religious prejudice."[23]

One can see how the early Calvinist conception that everything in one's life is driven by faith (and has religious significance) grew into the philosophical proposition that everyone is living by faith in some identifiable ideology. Dooyeweerd was arguing that "religious" (not necessarily deistic) faith in the form of pre-theoretical commitments play a role on the way to any stable knowledge. In this regard, Dooyeweerd can be accused of wanting it both ways. On the one hand, he wanted to be an ideological critic of modernity, insisting on the inevitability of belief structures;[24] on the other hand, Dooyeweerd cheerfully accepted the progress of science. In those regards, Dooyeweerd resembles Bruno Latour.

C. Parallels with Latour's two worlds

French (and Catholic) sociologist of science and technology Bruno Latour, in a move similar to Dooyeweerd's, is a famous critic of the scientific community's claim that its enterprise can somehow rise above cultural, linguistic, economic, ethical, and other social determinants:

> The ozone hole is too social and too narrated to be truly natural; the strategy of industrial firms and heads of state is too full of chemical reactions to be reduced to power and interest; the discourse of the exosphere is too real and too social to boil down to meaning effects. Is it our fault if the networks are simultaneously real, like nature, narrated, like discourse, and collective, like society?[25]

Latour, however, would *not* conclude that overt political interference with research (he references President Trump) is *proper*, and he has even recently acknowledged the reliability and necessity of the sciences for human progress and flourishing. This parallel with Dooyeweerd is not a mere coincidence; Latour's early social constructivism and later actor network theory remain as challenges to scientism, likewise a target of neo-Calvinist criticism. Yet, modern science was revered (some would say facilitated) in the Reformation, just as Latour reveres climate science in his recent *Down to Earth: Politics in the New Climatic Regime*.[26]

Dooyeweerd insisted that belief structures were inevitable, and yet he promoted consensus science. Latour is regularly accused of making exactly the same inconsistent move, that is, demonstrating that science relies on social, economic, and linguistic structures for its success, then introducing material nature (the "nonhuman") as the focus of and a limitation on scientific knowledge. The former made him a target of scientists accusing him of social constructivism— the postmodern threat to modern science—while the latter made him vulnerable to critique from his colleagues in the sociology of science and technology, who thought he was returning to a traditional idealization of science.[27] For Latour, however, science is a co-production of human actors and nonhuman *actants* in a network, and since science cannot "stand on its own," "[f]acts remain robust only when they are supported by a common culture, by institutions that can be trusted, by a more or less decent public life, by more or less reliable media."[28]

For Latour, even *artists* potentially provide support to the scientific enterprise because they are sensitive to, and can represent, the hard-to-capture complexities, novelties, and mysteries of science.[29] Art (including theatre, graphic novels, and painting) is one of Latour's three "aesthetics" (alongside science and politics) that can be mobilized to reveal the contours of the new climatic regime—not in the sense of simplistic, message-based ecological art, but, for example, to "dramatize and de-dramatize" the contradictions and divisions in our culture.[30] The primary "division" to which Latour refers is on the question of climate change. Its denial has resulted in the loss of a shared, *common* world—again, as quoted in Chapter 1, "there are now several worlds ... and they are mutually incompatible."[31] Recall here Kuyper's identification of two worldviews in conflict—Calvinism and modernism, both ideological—which is traceable back to Calvin's break from

Catholicism, a division that is also represented in de Witte's painting of the Old Church in Delft.

Latour shows that the traditional, idealistic image of a scientific fact as obviously true to everyone relied upon a framework of philosophical assumptions, experimental conventions, ethical beliefs, social interactions, heuristic metaphors, and financial resources. Science never was a matter of simply listening to Nature speak and recording the results, but it worked because of our common world. However, Latour observes, now "we have people who no longer share the idea that there is a common world. And that of course changes everything."[32] This two-worlds framework (obvious in the crisis of expertise during the COVID-19 pandemic) has implications for expertise in legal settings, whether in policy controversies or in the courtroom. Latour is quite clear that the Trump administration ignored the consensus science of government experts, particularly in the field of environmental regulation, where scientific decision making became politicized. Additionally, while Latour does not address expertise in the courtroom, the same problem persists when forensic science laboratories, idealized as "science," are on the side of, and controlled by, the police and prosecutors. The US National Academy of Science recently condemned the contextual bias in the supposedly scientific procedures of forensic scientists and called for independent forensic laboratories.[33]

The turn in the sociology of science and technology in recent years, exemplified in Latour's work, is toward defining and supporting consensus expertise in governmental settings, notwithstanding the discipline's previous emphasis on identifying social determinants in the scientific enterprise. In response to the criticism that sociologists of science and technology are now idealizing science, or that their previous constructivist *relativism* caused the politicization of science in policy contexts or the prosecutorial bias of forensic science, they would reply as Latour does: the sociology of science and technology was never a rejection of the best science that we have, and far from causing the loss of confidence in expertise, the current distortions of expertise demonstrate the validity of the concerns over social influences, some of which are inevitable, but some are problematic, like the influence of politics or prosecutorial bias on scientific findings.[34]

It bears mentioning that Harry Collins and Rob Evans, in their own turn to trust in consensus science—the so-called Third Wave of STS—have also faced criticism for emphasizing both the social aspects of expert communities and their engagement with natural reality. Critics of STS wonder (wrongly, in my view) whether the Third Wave is simply a return, in drag, to the scientistic idealism of the First Wave and, somewhat sneeringly, say, "Well, welcome!"—as if Collins and Evans have indeed returned to science's self-image as *not* co-produced by social structures.[35]

Nearly a century ago, Dooyeweerd was caught up in a similar controversy, *not* because he was a sociologist of science like Latour, visiting a laboratory to catalogue the social construction of facts, but because he was a devout Christian who would have appeared *biased* to secular scholars—he not only (audaciously) allowed his faith to influence his theorizing, but also claimed that such a framework of commitment was inevitable, whether acknowledged or not. Dooyeweerd made the argument, familiar in cultural studies and literary theory nowadays, that the autonomous Cartesian subject is a myth; the human subject is socially constructed in its early loyalties and dependence upon others—their images, their language, and their beliefs—for its identity, for its very *self*. Rawlsian public reason or common sense is therefore problematized, but that is not to say that everyone is robotic and predetermined. There is a middle ground, claimed by Dooyeweerd and Latour, where one need not decide between autonomous subjects producing neutral science and people with no choices who are irrational and doubt everything. Just as Latour fears the loss of a common culture in the post-truth era, dividing our society into two camps who do not live in the same world, Dutch neo-Calvinists like Dooyeweerd feared the marginalization of religion in the face of modern science, dividing our society into two camps, one of which saw religion (including scientism) as inevitable and the other who lived in a different world of presumably rational, Enlightenment subjects. The fictional conversation between two biochemistry graduate students reflects that division:

> "Do you believe in God?"
>> "Me? Well. No, I guess not. Not anymore."
>> "Science is kind of like God," Miller says.[36]

Scientism is not unlike religion.

The 16th-century iconoclasm in the north of Holland was not just about theology (for example, "only faith, not works," "only scripture, not churches," "only Christ, not priests"). The rejection of Catholicism, obvious from the removal of images by whitewashing church interiors, actually went farther to claim a new worldview, a new ideology, and, in Latourian terms, a new world in competition with that of the papacy. While the strong disagreements that characterized the Protestant Reformation were never resolved and therefore do not provide a roadmap for overcoming the crisis of expertise, they nevertheless suggest the inevitability of ideological commitments in public policy disputes involving scientific matters.

THREE

The Quasi-Religious Aspect of the Crisis

A. Tribalism in politics

We tend to think of political divisions primarily as "secular"—for example, in the US, even when one major political party becomes associated with evangelical Christianity, some supporters of that party will not share those beliefs and members of the opposition party may reject religion altogether (identifying as nonbelievers) or claim that religion should not influence politics. However, that limited understanding of religion is misleading: "Whereas the foundational metaphor for tribalism is kinship, the foundational metaphor for political sectarianism is religion, which evokes analogies focusing less on genetic relatedness than on strong faith in the moral correctness and superiority of one's sect."[1] A recent *Science* article on political partisanship by social psychologist Eli Finkel (Northwestern University) and others analyzes our cultural divide in terms of opposing sectarian faiths.[2] Mirroring the Protestant Reformation in Holland discussed in Chapter 2, the core ingredients of political sectarianism identified by the authors are: (1) "othering," in the sense of exaggerating differences; (2) strong aversions to, even contempt for, those in the other group; and (3) moralization, as the other side is characterized as disgraceful or iniquitous due to belief in the moral superiority of one's community.[3] Despite "plentiful" common ground, the growing, mutual distrust nowadays in the US between Republicans and Democrats leads to a striking polarization.[4] In a manner reflected in religious sectarianism, "Americans today are much more opposed to dating or marrying an opposing partisan; they are also wary of living near or working for one."[5] We can discern "radically different sectarian narratives about American society and politics."[6]

Like other analyses of the culture wars, the authors note that "as Americans have grown more receptive to consuming information slanted through a partisan lens, the media ecosystem has inflamed political sectarianism."[7] Like the analyses of the crisis of expertise by sociologists, the authors also recommend instilling into political debates:

> intellectual humility, such as by asking people to explain policy preferences at a mechanistic level—for example, why do they favor their position on a national flat tax or on carbon emissions? According to a recent study … those asked to provide mechanistic explanations gain appreciation for the complexities involved. Leaders of civic, religious, and media organizations committed to bridging divides can look to such strategies to reduce intellectual self-righteousness that can contribute to political sectarianism.[8]

It should be noted that the foregoing analysis is not one-sided, accusing the Right of othering, aversion, and moralization; rather, the implication is that both sides exhibit sectarian impulses. In the end, Finkel and his co-authors turn away from the analogy with religion: "In the United States today, even though Democrats and Republicans differ on average in terms of religious affiliation, their schism is fundamentally political rather than religious. It is, in this sense, quite distinct from the Sunni-versus-Shia sectarian schisms that characterize politics in some Muslim-majority nations."[9] I disagree, only because in this book, I am neither characterizing "religion" as belief in a deity, nor limiting the analogy with religion to the situation where the Right and the Left each have their own deistic religion (they do not, of course). I am using the term "religion" in the sense of having faith-like commitments to a set of beliefs and values.

B. Anti-science ideology?

Professor Shi-Ling Hsu (Florida State University) offers an analysis of the culture wars, the crisis of expertise, and our current political polarization in terms of ideology—a seemingly similar approach to my own. Hsu identifies on the part of Trump supporters "an *ideology* of hostility to science," representing a new episode in the history of populist anti-intellectualism.[10] He also acknowledges the history, long before the Trump presidency, of attempts to obstruct science to avoid "pesky regulation" of, for example, "the fossil fuel and chemical industries."[11] Hsu identifies, however, a *new* attack on scientific research itself, and it is that phenomenon that makes "skepticism and hostility to science into an *ideology*."[12] Hostility toward scientific *experts* is also evident: "large blocks of voters apparently are willing to believe that scientific experts might be part of a 'mainstream establishment' conspiring to oppress them."[13] In his introduction, Hsu even tries to see both sides in the culture wars as ideological: "[H]ostility or distrust of science is certainly not limited to the Republican Party or those on the political right. ... [S]ome people, predominately on the political left, have continued to abstain from consumption of [GMO foods]. Populist suspicion of vaccines is persistent ... and bipartisan."[14] Yet, other than possibly the proponents of the Green New Deal, Hsu argues that "left-wing suspicions" about science "have not metastasized into cultural identifiers."[15] Hsu's critique of ideology, therefore, from his (1) summation of the history of anti-intellectualism (only recently focused on the physical and biological sciences), to (2) the phenomenon of climate-change denial, to (3) the Trump administration's political interference with science (including disbanding scientific advisory panels, doctoring cost–benefit analyses to help industry, and relocating and "hollowing out" the Economic Research Service), lands squarely on those conservatives who seem to want "to have government by ideology alone, devoid of science."[16]

Anti-science ideology works, Hsu explains, in part, because scientific experts are seen as part of "the establishment" or "the deep state"—our "dependence upon a vast network of government experts breeds suspicion and resentment," especially when those experts are viewed as a privileged elite (with powerful knowledge most cannot understand).[17] Falsely linking "job losses to science-backed environmental regulation" also helps fuel anti-science ideology, as does the fate of white working-class Americans:

> Finding themselves in opposition to a panoply of non-white, non-Christian, non-heterosexual groups, [Amy] Chua argues that the newly impoverished white working class seeks desperately to coalesce to regain political power they perceive they have lost. It is not hard to see how part of that white working class identity, rooted in grievance, would find scientific experts, including economists, to be part of the despicable "other."[18]

Add to this the ubiquity of experts during the COVID-19 pandemic, the collapse of professional journalism ("losing out in competition to social media"), and the way "Trump and kindred Republicans are trafficking in misinformation about hard, provable scientific facts that are susceptible of empirical verification,"[19] and you have, in Hsu's view, anti-science politics.

This view finds support in the former Trump administration's pattern of disrupting our "usual scientific processes," which was due to that administration's fear that development and expression of genuine "scientific evidence or insight might threaten achievement of major ideological or political objectives."[20] For example, a study on the health effects of mountaintop-removal coal mining was halted, while the Trump administration's Environmental Protection Agency (EPA) administrator "asserted the fringe view that ... carbon dioxide is not a primary contributor to global warming."[21] Such pseudoscience can become "a cultural force" adding to the mistrust of scientific institutions:

> [pseudoscientists] argue that ... scientific consensus emerges from a conspiracy to suppress dissenting views. They produce fake experts [with] views contrary to established knowledge but [with no] credible scientific track record. They cherry-pick the ... papers that challenge the dominant view as a means of discrediting an entire field. ... And they set impossible expectations of research.[22]

In short, pseudoscience has the "form" of science without the substance.[23]

Hsu is to be commended for carefully describing the bubble (of hostility toward science) occupied by the Right. Despite his effort to be fair, however, the analysis stops short of considering whether the Left has its own bubble—its own assumptions, interpretive lenses, values, and cultural identifiers, in short, its own ideology. That is why, in Chapter 4, I adopt a view of ideology as an inevitable quasi-religion made up of believed-in values, identities, and priorities. The humility recommended by many analysts of the *political* culture wars is clearly based on the idea that we all (Left and Right) have precommitments and values that affect our respective viewpoints, and the humility recommended by analysts of the crisis of expertise is based on the idea that science is not a perfect producer of unmediated facts. Perhaps, as I explain in Chapter 4, it would be better to view each side's respective beliefs about all matters scientific, whether consensus or fringe, as a form of *expertise*.

FOUR

Belief as a Form of Expertise

Anthropology is famous for its effort to understand a culture other than the anthropologist's own—its ethnographic methodology in particular aims to see things as others do: "What they call life is a ghost ship. On the ship are many rooms. ... An uncountable number. There is always another. This is how they escape the prison of the self. To see the world through the windows of someone else's room."[1] Impliedly, we all have a perspective, or a viewpoint from which we see the world; the challenge in the crisis of expertise is to understand why others might disagree about something as seemingly universal and uncontroversial as scientific facts.

A. The anthropology of religion

Anthropologist Tanya Marie Luhrmann (Stanford University) recently published a study of religious *practices*—not simply religious *beliefs*—entitled *How God Becomes Real: Kindling the Presence of Invisible Others*.[2] Luhrmann emphasizes the "real-making" efforts, including prayer, ritual, and worship, of religious believers. Her ethnographic observations, for example, of charismatic Protestants "suggested that it took these staunch evangelicals [a lot of] effort to keep God present and salient in their lives."[3] Luhrmann recalls public debates between atheists and Christians, the former convinced that the way Christians "think is simply wrong-headed," and the latter explaining that "Christianity is not about propositions at all, but rather about truths that are more transcendent, symbolic, and nonliteral":[4] " 'The result,' [anthropologist Jonathan] Mair comments, 'is a loud conversation at cross purposes.' That's because, he argues, they think about realness differently. The two sides don't hear each other properly because they live in different 'cultures of belief.'"[5] Even the atheist, that is, lives in a culture of belief. Both have a sustained, intentional, deliberative commitment: for the Christian, it is a faith in "the idea that there are invisible beings who are involved in human lives in helpful ways";[6] while for the atheist, it is a firm belief that there are no such beings. And both have a set of values to which they are committed.

In faith communities, Luhrmann also identifies how religious people occupy a "special world defined by special rules. ... [T]hey signal their participation by special actions [called] rituals ... [and] find their inner lives socialized by others as they accept the rules of the game and remake them as their own."[7] Religious commitment, far beyond a mere passive set of beliefs, involves "narrative and practice," absorption ("once you get absorbed in something, it seems more real to you"), "cultivation of inner imagery," and even training ("to *learn* to hear God speak").[8] There is faith—the "beliefs, thoughts, and attitudes"—but there is also "the way the feeling of realness is kindled through practices, orientations, and the training of attention."[9] This distinction between beliefs and *practice* is perhaps made clearer when one recalls how evangelical Christians pejoratively identify, in mainstream Protestant churches, "nominal Christians"; the implication is that the faith of such liberal Christians-in-name-only does not really have any practical effect on, or cause changes in, their lives.

Luhrmann does not use the word "expertise" to describe these religious experiences and practices, but James Wood, her reviewer in *The New Yorker*, does:

Hearing God's voice, [Luhrmann] says, is "richly layered skill," and her subjects speak of developing it as one would speak of any expertise; they think that "repeated exposure and attention, coupled with specific training, helps the expert to see things that are really present but that the raw observer just cannot."[10]

Luhrmann's references to "training and technique," and to practices that "work," or that "change people" and "change mental experience," appear to justify the analogy with expertise.[11] In the end, however, Woods back off and seems to doubt there can be expertise when we are dealing with (what Wood thinks is) an illusion: "Wine tasters and sonogram technicians are drinking real wine and looking for real babies, but ghostbusters and psychics might be thought to have 'trained' themselves to find mere figments."[12] I think Wood is wrong here, not necessarily about his implied proposition that ghosts are an illusion, but about whether there can be expertise in things like witchcraft ("Like evangelicals … witches stud[y] how to wield their specific magical powers"[13]). Expertise, in the view of some who carefully study that phenomenon, does not need to correspond with a judgment of correctness, truth, or reality.[14] I am referring here to the relatively recent movement in the sociology of science known as "studies of expertise and experience," also called the "Third Wave in STS."[15]

B. Studies in expertise and experience

For over 15 years, sociologists of science Harry Collins and Robert Evans have been developing an analysis of expertise based, in part: (1) on philosopher Peter Winch's *The Idea of a Social Science*,[16] which discussed "the implications of [Ludwig] Wittgenstein's later philosophy for sociology";[17] and, consequently, (2) on Wittgenstein's concept of *Lebensform*: "[Collins and Evans's] approach takes it that there are 'forms-of-life' (cultures or paradigms) characterized by certain ways of going on and ways of thinking and that those who are fluent in these ways of going on and thinking are experts in those domains."[18] Here is the analogy with religion in Luhrmann's terms: Collins and Evans argue that to "become an expert in some domain is a matter of becoming embedded in the social life of the domain."[19] Expertise, in this view, is limited neither to (1) those with "more true and justified beliefs than experts," nor to (2) those with "hard-won" esoteric skills or knowledge—one can be an expert in the complexities of linguistics, or merely in speaking English.[20]

My proposal to label as "experts" not only (1) consensus scientists, but also (2) marginalized and minority-view scientists, raises a mistaken concern that I am equating the usefulness or helpfulness of these two communities. First, as discussed throughout this book, expertise does not imply correctness (hence my similar labeling as experts both [3] non-scientist believers in consensus scientists and [4] non-scientist believers in fringe science). Second, there is a paradoxical benefit to recognizing fringe science as "science" even as it is based on questionable data and is eventually discredited—what Eyal calls a "recursive dynamic": "In controversies about climate change, for example, while 'scientists are treated as proxies for interest groups,' *science* itself 'is held up as universal and impartial'."[21] Eyal's point is that attacks on consensus science, reflecting distrust of mainstream science, are often assaults not on *science itself*, but on a particular position in a controversy. That is why critics of excessive government regulation of industry who are "demanding rigorous proof" (for example, that an industry is polluting) end up arguing that "more science is needed."[22] My point, while agreeing with Eyal, is that the consensus and fringe scientific communities are both claiming expertise in some field of *scientific* inquiry; they are not properly seen, respectively, as scientists and anti-scientists—although those who idealize consensus science, unhelpfully in my view, might disagree. Both communities are, instead, *forms of life*.

C. Wittgenstein's influential notion of "forms of life"

Wittgenstein, like any philosopher, is not always easy to understand, especially when we are asked to understand a new concept: "[T]o imagine a language means to imagine a form of life. ... [T]he term 'language-*game*' is meant to bring into prominence the fact that the *speaking* of language is part of an activity, or of a form of life."[23] Commentators, also philosophers, have not always agreed on exactly what Wittgenstein means:

> Wittgenstein's use of the expression "Lebensform" has occasioned much controversy. ... It has a long history prior to his employment of the expression. ... [It] occurs in the *Philosophical Investigations* only three times. ... [H]owever, the general conception that underlies this very sparse invocation of the expression, and its links with other notions in Wittgenstein's later philosophy and methodology ... are of capital importance.[24]

The numerous interpretations of Wittgenstein's term "forms of life" have engendered a substantial literature on the topic. Is a "form of life" (1) simply a language-*game*—another of Wittgenstein's key concepts—insofar as "forms of life are shared and standardized"?[25] Is it (2) a tendency "to behave in certain ways" (a concept "connect[ed] so closely to language-games that a question arises whether it is useful to distinguish them")?[26] Or, is it (3) "a way of life, or a mode, manner, fashion, or style of life [having] something important to do with class structure, the values, the religion, the types of industry and commerce that characterize a group of people"?[27] In J.F.M. Hunter's view, a better interpretation is to see a "form of life" as (4) "something typical of a living being," a biological or organic phenomenon.[28] At the risk of not fully appreciating Hunter's sophisticated reading of Wittgenstein and defense of the organic account, it is worth noting that a substantial controversy persists regarding whether Wittgenstein's use of the term "*Lebensform*" is indeed offering a biological account of a single human form of life (Hunter's interpretation) or, rather, an explanation of the *historico-cultural* differences between multiple forms of life, roughly the third interpretation earlier.[29] In the former (the biological) interpretation, a form of life is universal, namely, the "common behavior of mankind," whereas in the latter (the historico-cultural interpretation), "there is only a plurality of forms of human life,"[30] each one a unique "way of living, a pattern of activities ... inextricably interwoven with, and partly constituted by, uses of language."[31] Both interpretations are likely justified:

> Whereas all humans share in a fundamental form of life, there [also] exist ... patterns of life—possibilities for diversity and variation[, that is,] for ... various forms of human life. ... So that where the acquisition of language belongs to the human form of life, the acquisition of cartography, or of algebra, or of parliamentary elections attaches only to some of the various forms of human life.[32]

We can see an example of a form of life, according to Stanley Cavell, in those who are religious.[33] Yet, to "understand ... an utterance religiously you have to be able to share its perspective"[34]—this "is not to say that we must *belong* to a form of human life in order to understand it, but it does mean that we must be able to share its perspective."[35] That distinction between belonging to a community and "sharing" its perspective is important for understanding Wittgenstein and for explaining how the opposing sides in the crisis of expertise might understand, communicate with, and respect each other more.

FIVE

Communicating across Worldviews

A. Wittgenstein's hope

Even as we seem to speak the same language as our fellow citizens, often we do not: "[W]hile we might understand that there are many peoples speaking many different languages, we are fooled into thinking that everyone in our own tribe speaks the same language we do."[1] However, one of the insights from Third Wave studies of expertise and experience is that we do not have to become *believing* members of another community to learn its language: "[I]t is possible, given the right circumstances, for a competent human from any human group to understand the culture of any other human group without engaging in their practices."[2] Moreover, there is in Wittgenstein an optimism that it is possible for those who occupy a form of life to communicate with, and even persuade, those in a *different* form of life. The philosophical problem of relativism, however, arises in Wittgenstein's account of forms of life—both *cultural* relativism due to differences between groups "with regards to social, moral and religious values and practices," and *cognitive* relativism due to differences between groups as to their "different ways of 'seeing the world'— that is, that there seems to be a plurality of different sets of categories under which experience is organized and the world understood."[3] If others live in a different world to me, or see the world differently than I do, how can I even talk about what is wrong and right about current events, appropriate values, or scientific knowledge? For Wittgenstein, however, we can "imagine situations and practices that are quite different from our own. ... Outsiders can, as it were, achieve something of an insider's perspective. ... [D]ifferent ways of 'seeing' the world are not cognitively inaccessible to one another."[4] For example, Wittgenstein suggests that we should try to *persuade* a person who believed the earth to be only 50 years old: "We should be trying to give him our picture of the world."[5] It is never, however, merely "a matter of presenting cold facts about [his] false beliefs."[6] In Philip Toner's formulation, "[d]ialogue, persuasion, self-awareness and humility are the order of the day for Wittgenstein."[7]

Finally, the literature addressing Wittgenstein's notion of forms of life also includes a discussion of religion—about whether a religion is a *form of life*—including Wittgenstein's own comments on religious belief. Even though that discussion is primarily about deistic religions, it is relevant to this book and my own use of the term "religion" to denote a quasi-religion that may or may not involve fundamental beliefs or rituals concerning a deity (or deities). There is an important suggestion in Wittgenstein that a religious community is a form of life.[8] Alan Keightley, in *Wittgenstein, Grammar and God*, after acknowledging that the "precise meaning" of forms of life (or of *language-games*) "is the source of great confusion," points out that Wittgenstein rejected ethnocentrism in the anthropology of primitive religious ceremonies:

> The practices are not about bad scientific theories, and the participants are not "stupid." The ceremonies only appear superstitious if we neglect the role they play in people's lives. ... Even where religions express their differing traditions (e.g., Augustine and a Buddhist holy man), it does not follow that some of them must be mistakes.[9]

A mistake, for Wittgenstein, implies a "context of opinions, hypotheses, and explanations"—we should "resist the tendency to explain beliefs and practices" because religious "beliefs are *absolutes*" that reveal their nature "by the way in which [they regulate] all of the believer's life."[10] Religious "beliefs are not *true* in the sense that they are propositions based on good evidence"—to characterize a disagreement between a believer and an unbeliever as a division "about the reliability of certain evidence" is to misunderstand "the nature of the difference between them."[11] The two sides "think differently, in a different way":[12]

Expressions of belief are not used in the same way as hypotheses in science or history. ... Religious truth ... misconceived [as a scientific proposition], appears as a "blunder" out of its "home" system. It strays into a foreign system or "game" where scientific reasonableness rules. ... To speak of the "truth" of religious beliefs, then, is to say that one lives by them, that they constitute the framework for one's life.[13]

If religious beliefs are not hypotheses, but expressions of faith and trust, it is easy to see that they are different from the practice of science (indeed, they are unlike "the whole spectrum of social life" where we form and test hypotheses[14]). Even so, we should not confuse the practice of science with factuality. In Karl Popper's idealized view of "the laws of nature which are established and presupposed in all the sciences" ("Facts are *there*"), as distinct from human decisions and behavior, he "ignores the way in which the formulation of laws of nature ... emerge from particular societies."[15] This is where we return to Winch's view of factuality, inspired by Wittgenstein:

[W]e cannot say without qualification that modern scientific theories are about the same set of facts as the theories of earlier stages of scientific development were about. ... [S]cientific concepts have changed ... and with them scientists' view on what is to count as a relevant fact. ... [T]his does not mean that earlier scientists had a *wrong* idea about what the facts were; they had the idea appropriate to the investigations they were conducting.[16]

As Thomas Kuhn argued in his paradigm theory (*after* Winch published *The Idea of a Social Science*), there "cannot be a simple notion of 'the facts' which is not itself 'theory laden.'"[17] This is the basis for the modest view of the scientific enterprise, represented in the work of sociologists of science, often called the "sociology of scientific knowledge," or "science and technology studies." Science is not simply a matter of letting Nature speak, but rather a sociocultural activity with linguistic, institutional, and economic aspects. Collins and Evans's approach in their "studies of expertise and experience," even as it emphasizes that consensus science approaches truth and should be at least tentatively accepted as authoritative, arises out of and maintains the essence of that tradition.

For my purposes in this book, I want to emphasize three developments by Collins and Evans derived from Winch and from their engagement with Wittgenstein's forms of life. First, a domain of expertise is a group of specialists who have "a common set of concepts and a common spoken discourse—the 'practice language' of" that domain.[18] Second, it is possible to communicate with those specialists by understanding their language, and therefore their practical world, even if one does not practice their specialization.[19] Finally, third, this ability to communicate introduces the concept of *interactional experts*. It is possible for those who are not trained or experienced in a domain of expertise to learn the language of the *contributory experts* in the domain—the core group who are able, based on their training or experience, to contribute to their field—and interact with them, and to thereby develop expertise. The easiest example of these two forms of expertise—contributory and interactional—is in the context of an esoteric scientific discipline, such as gravitational wave theory: some are trained, experienced, and published in the relevant science (thus contributory experts); while others might embed themselves among those scientists, learn their language, and perhaps even "imitate" them such that they could be mistaken for someone who practices the science.[20] However, to the extent that the domain of expertise is not esoteric, for example, those non-scientists who believe and support the minority of scientists who question mask wearing or the efficacy of vaccines, then the contributory experts would be the non-scientist members of that community who practice those beliefs (perhaps by not wearing a mask or by refusing a vaccine) and an interactional expert might be a non-scientist believer in *consensus* views on the benefits of mask wearing or vaccinations. That non-scientist interactional expert who believes in consensus science may desire to understand the practices and language of the community of those who do not trust consensus science to communicate with them and perhaps persuade them to change their views.

B. Science communication studies

The difficulties of communicating the consensus view of a majority of scientists took on a new importance during the COVID-19 pandemic:

> Historically, the two most effective responses to vaccine skepticism have been government mandates and relentless, calm persuasion. But broad Covid-vaccine mandates are probably unrealistic in the U.S. today, thanks to a combination of a Supreme Court ruling and widescale public opposition. Persuasion will probably have to do most of the work.[21]

There can be little doubt that science communication, as a cultural phenomenon and as a field of study, is important: citizens and politicians need high-quality scientific knowledge to live their lives and to govern.[22] However:

communicating science is almost always a complex task in part because scientific information and its implications are understood, perceived, and interpreted differently by different individuals. ... In communicating about science-related controversies, such factors as conflicting values, competing economic and personal interests, and group or organizational loyalties can become central to a person's individual decision.[23]

STS scholars therefore regularly reject the "deficit model" of science communication, whereby "scientific knowledge and appreciation can be injected into members of the public"[24]—announcing consensus is not the same as understanding public concerns.

Arlie Hochschild's concept of "deep stories," developed in her journey through right-wing US communities (including "Tea Party" adherents, mainly white, Christian Republicans),[25] captures the set of central values and interests that constitute reality for those with flat wages and job insecurity. They feel betrayed by, and distrust, a welfare state that unfairly allows immigrants to take their jobs or minority groups to get ahead of them (that is, "line cutters" who benefit from affirmative action).[26] This worldview—"political feeling runs deeper than it did in the past"[27]—is the same one that might resist environmental regulation on economic grounds ("politics is the single biggest factor determining views on climate change"[28]), or vaccination mandates nowadays on the grounds of personal freedom and distrust of government experts. The more important point in this chapter, however, is Hochschild's experience in "reaching out to someone from another world, and of *having that interest welcomed*"; instead of easily settling "for dislike and contempt,"[29] which is so easy in our polarized culture, she took the time to get to know the subjects of her research in light of the increasing reality that we simply do not know each other.

Given that "people tend not to adopt explanations that conflict with their long-held views or values," one of the strategies "that can be used to mitigate the effects of competing beliefs, values, and interests on science communication [is to tailor] messages from science for understanding and persuasion ... while still offering accurate information."[30] Notably, in light of my thesis concerning the inevitability of worldviews, scientists are advised "to avoid conflating scientific understandings with their own values. ... They cannot assume that their information conveys a moral imperative or presume that their own values are universal."[31] This raises the question of whether scientists who become political activists will lose credibility. Heather Douglas, in her important work on the inevitability of values in scientific research, suggests that the boundary between science and policy is not so clear (even though she rejects "politicized" science that lacks intellectual integrity)—a view that seems to encourage political activism.[32] Collins and Evans warn against scientists' tailoring their views to selected "citizens and stakeholders"—that "is not what scientists do"—even though "no one can be against the softening of boundaries between the scientific profession and society as a whole."[33] Latour, on the other hand, seems to encourage activism on the part of experts:

> Climatologists ... must recognize that, as nature's designated representatives, they have always been political actors, and that they are now combatants in a war whose outcome will have planetary ramifications. We would be in a much better situation [if scientists] stopped pretending that "the others"—the climate-change deniers—"are the ones engaged in politics and that you are engaged 'only in science.'"[34]

Notwithstanding his use of military metaphors, Latour is sensitive, in terms of science communication, to the need for a "greater understanding of the circumstances out of which misinformation arises and the communities in which it takes root," which "will better equip us to combat it."[35]

While persuasive marketing has a bad reputation for selling unnecessary products to consumers, there should be less concern with persuading citizens "to change their attitudes or perceptions or to take action for the public good based on established scientific evidence."[36] Another recommendation for scientists is to concede that uncertainty is inherent in science, projecting modesty, even as they "communicate repeatedly the extent of expert agreement on the science concerning a contentious issue."[37] However, as to scientists who "feel an urgent need to correct information that is inconsistent with the weight of scientific evidence," that can be problematic: "when people are challenged in their beliefs, they may react by dismissing the credibility of the messengers who provide the corrections."[38] Finally, public participation in formal decision-making processes can improve communication of scientific knowledge:

> When done well, public participation improves the quality and legitimacy of a decision and builds the capacity of all involved to engage in the policy process. It can lead to better results in terms of environmental quality and other social objectives. It also can enhance trust and understanding among ... stakeholders [and] can be effective in diminishing controversy around science in the public sphere.[39]

It is here, on this recommendation of public *engagement* in scientific controversies, that my approach is distinguished from the views of policy analysts like Frank Fischer, discussed in the next section. Clearly explaining a scientific controversy to citizens and even letting citizens "participate" by collecting data are obviously beneficial in the effort to overcome the crisis of expertise. In my view, however, with respect to notions of citizen assemblies and epistemic democracy, grounding scientific expertise in lay communities, as opposed to trained scientists or experts through experience (and thus not "lay"), is highly problematic.

C. Deliberative approaches and the "argumentative" turn

Before distinguishing my proposal (for overcoming or at least minimizing the effects of the crisis of expertise) from Frank Fischer's "argumentative" or deliberative approach to policy disputes, he should be credited as the unofficial surveyor of the many analytical frameworks that reject the just-announce-consensus-science model in the contemporary context of tribal political divisions, fake news, and distrust of governmental technocrats. Fischer includes in the "argumentative turn" the various emphases, by *early* commentators, on "practical argumentation, policy judgment, frame analysis, narrative storytelling, and rhetorical analysis."[40] The argumentative turn, Fischer continues, has now expanded "to include work on discourse analysis, deliberation, deliberative democracy, citizen juries, governance, expertise, participatory inquiry, local and tacit knowledge, collaborative planning, the uses and role of media, and interpretive methods, among others."[41] While I find the emphases on expertise, discourse analysis, communication and argumentation compelling, it is the inclusion of the popular notion of *citizen juries*, as well as *collaborative* planning and *participatory* inquiry, to the extent that these approaches go beyond understanding and listening to stakeholders, that I find troubling: decisions on scientific matters should not, in my view, be made by ordinary citizens unless they have either training or experience in the relevant field (in which case, they are not ordinary citizens).

That said, Fischer's identification of argumentation with deliberation and rhetoric is, to me, the key to "civil debate" and "persuasive dialogue."[42] In the ideal policy situation to which we should aspire, "[T]he arguer pays special attention to those he or she is speaking to, their beliefs, backgrounds, intellectual styles, and communicative strategies. ... The arguer attempts to persuade the audience to see and understand something ... one way as opposed to another."[43] Communication "within and across discourses"[44] is, at least in this formulation by Fischer, a matter of convincing someone to accept, for example, scientific consensus on some contested matter. The crisis of expertise has demonstrated that ensuring public confidence in consensus science is never just a matter of an assertive technocrat—of "replac[ing] the talk of politicians with the analysis of experts"[45]—but rather an exercise in modesty and persuasion. The goal, in my view, is bolstering support for consensus science, *not* ensuring that misinformed and doubtful minority views are left alone without criticism in the public square.

Related to my concern over Fischer's emphasis on "democratic science," I also want to interrogate briefly the notion that policy science is improved by "competing perspectives on the part of both government officials and public citizens."[46] In the crisis of expertise, competing viewpoints are more of a problem than a solution. That statement may distinguish my approach from that of Naomi Oreskes in *Why Trust Science?*, who likewise emphasizes a diversity of viewpoints as a key to trustworthy science: "[O]bjectivity is likely to be maximized when there are recognized and robust avenues for criticism, such as peer review, when the community is open, non-defensive, and responsive to criticism, and when the community is sufficiently diverse that a broad range of views can be developed, heard, and appropriately considered."[47] I agree completely with Oreskes's recommendation that we should trust *consensus* science because of its method and evidence, even as she properly, in my view, accentuates its social character, its imperfections, and its uncertainties.[48] Indeed, the occasional arguments *against* trusting consensus science—including the marginalization of dissent, announcing consensus too soon, or overuse of the locution "science says"[49]—are actually a recognition that we should always acknowledge uncertainty in science. I also agree that a broad range of views should be heard and *appropriately* considered, with the hope that "appropriately" includes, for Oreskes, respectfully persuading those with fringe scientific views to change their beliefs. It is a subtle point because I can certainly buy into demographic (for example, race and gender) diversity in science, and even "perspectival" diversity[50]—after all, I have been arguing that those with opposing worldviews should respectfully communicate with each other regarding their respective scientific arguments. I agree that a scientific "community with diverse values is more likely to identify and challenge" embedded beliefs[51]—hence, I have been arguing that no one should think they can rise above ideology. However, the notion of epistemic strength in diversity sometimes sounds too much to me like citizen science—giving a decisive voice to ordinary non-scientists in policy decisions—which will neither increase public trust in consensus science, nor help resolve the crisis of expertise.

Conclusion

Throughout this book, I have maintained both that ideology is inevitable and, consequently, that we cannot criticize others from a position of neutrality: "For Latour the critic pretends to an enlightened knowledge that allows him to demystify the fetishistic belief of naïve others. … [T]he fatal mistake of the critic is not to turn this anti-fetishistic gaze on his own belief … a mistake that renders him the most naïve of all."[1] Well before the crisis of expertise was intensified, due, in part, to the Trump administration's response to global warming and the COVID-19 pandemic, Latour explained where such naivete leads: "This is why you can be at once and without even sensing any contradiction … an antifetishist for everything you don't believe in … and … a perfectly healthy sturdy realist for what you really cherish."[2]

The situation with respect to consensus science therefore seems dire, as citizens living in different worlds cannot believe how blind their opponents are. Collins and Evans, however, in their analysis of communication between experts, rely on a Wittgensteinian view of lived worlds, wherein communication, in the sense of understanding the other side, is possible even between opposing worldviews. There may be limitations in the culture wars, insofar as a journalist who tries to embed within a far-right militia group in order to understand its beliefs might find that assignment difficult. However, in the field of scientific expertise, it is not difficult to imagine taking seriously (and respectfully criticizing) some of the scientific claims and data of minority-view scientists, all the while acknowledging the uncertainties and limitations of consensus science. With respect to controversial scientific issues, it is rarely a matter of claiming the unadorned "Truth" and identifying the stupid people who will not acknowledge it. We need, that is, to recognize the tension "between the need to preserve the right of the individual to make novel claims, setting him- or herself outside of the consensus, and the need to accept a degree of regulation of scientific thinking and acting if science is to move forward. … [S]cience is always balancing these two needs."[3] My conclusion is to call for humility with respect to our scientific beliefs and to claim that we can understand those in other bubbles and communicate persuasively. I recognize that tribal divisions in a polarized society make this task difficult, but it must be pursued.

Right after the insurrection at the US Capitol on January 6, 2021, Reverend Planning offered a Jesuit perspective on the culture wars and the possible solution:

> St Ignatius lived during the Protestant Reformation. It was a time of clashing worldviews not unlike our own. … Ignatius suggested that we always try to put the most positive interpretation on the views of a person who thinks differently from us.[4]

That may be idealistic, and probably *will not* work for the US culture wars generally (Will Democrats put a positive spin on the suggestion that the election was stolen?), but for the crisis of expertise, there is hope in the possibility that one can at least see and acknowledge the *scientific* evidence—albeit evidence that consensus scientists find insufficient, outdated, or misleading—in minority-view science. Just as in "the sociology of scientific knowledge … you are always trying to enter the [scientific] other's domain," so those who engage, and try to convince, "experts" with opposing views need to be able to "step from one worldview into another."[5]

Finally, it bears mention that some have blamed the discipline of the sociology of science, a field known for its modest view of the scientific enterprise, for causing distrust of science on the part of many citizens.[6] While it is true that the sociology of science, with its emphasis on the social, rhetorical, and economic aspects of science, presents a challenge to idealized views of science as a matter of listening to Nature speak,[7] it is hardly the case that ordinary citizens are familiar with the literature in that specialized field. Nevertheless, in a curious reversal, far from being the cause of distrust in science, the modest view of science is likely the solution to the crisis of expertise.

Journalist Nicholas Kristof spoke to black musician Daryl Davis about how to defeat the racism Davis experienced as a youngster:

> One of Davis's methods—and there's research from social psychology to confirm the effectiveness of this approach—is not to confront antagonists and denounce their bigotry but rather to start in listening mode. Once people feel they are being listened to, it is easier to plant a seed of doubt. … [This] approach is out of step with modern sensibilities … [for example, the] impulse … to decry from a distance. … [However,] "you won't get through to people until you've earned their trust."[8]

This is not a quick solution, but rather a long-term project— "[o]ne conversation at a time," Kristof says.[9] That realization mirrors Gayatri Chakravorty Spivak's rough analogy of political practice to *housework*, insofar as you "don't do [things like housework or personal hygiene] once and for all": "[W]ho doesn't know this? Except political theorists who are opining from the academy with theological solutions once and for all. I mean, political practice is more complex than housework, but … it involves the same persistent effort."[10] Likewise, the efforts recommended in the book—toward understanding another's perspective and the values that might drive their reactions to consensus science, demonstrating modesty with respect to that scientific consensus, and persuasion of those enamored of fringe or minority scientific views—will need to take place on a daily, persistent basis.

Notes

Preface

[1]   Wittgenstein (1980: 9e).

Introduction

[1]   Regarding the scientific consensus on human-caused global warming, see Cook (2016). Market research, however, shows that many voters "believe that there is no consensus about global warming in the scientific community" (Luntz, 2002, quoted in Cook, 2016).

[2]   Concerns over vaccines causing autism were legitimized by a 1998 study by Dr Andrew Wakefield in *The Lancet* medical journal, but that article was retracted on February 2, 2010, due to accusations of unethical and irresponsible research (see Hadhazy, 2010). Recently, there are concerns, likely unjustified, about the dangers of COVID-19 vaccines (see McLaughlin and Dzhanova, 2020).

[3]   Some argue that the term "filter bubble"—a state of intellectual isolation brought on by website algorithms that filter out disagreeable information—is an advance over the term "echo chamber" (see Pariser, 2011):

> In the filter bubble, there's less room for the chance encounters that bring insight and learning. ... By definition, a world constructed from the familiar is a world in which there's nothing to learn. If personalization [via filters] is too acute, it could prevent us from coming into contact with the ... preconception-shattering experiences and ideas that change how we think about the world and ourselves. (Pariser, 2011: 13)

Moreover, when a filter-bubble occupant does confront an opposing perspective, logical arguments may not sound compelling due to identity politics: "[P]olitics is not just about making the most logical argument. It also needs to be appealing to the imagination and identity of the people it concerns, and is often a case of trying to convince people 'who we are' in terms of shared identity and values" (Gilroy-Ware, 2020: 19).

[4]   See Richard (2020). The idea that those who differ politically are living in different universes, or "alternative worlds," is common in contemporary

discourse: "[Newt] Gingrich ... said ... liberals and conservatives 'live in alternative worlds. You have more than 74 million voters who supported President Trump despite everything. ... The truth is tens of millions of Americans are deeply alienated and angry'" (Gingrich, *Washington Post* op-ed, quoted in Mastrangelo, 2020).

5   See, for example, Hsu (2021).

6   See, for example, Gilroy-Ware (2020: 5): "[T]echnology platforms [enable] misinformation and disinformation." See also Hsu (2021: 456): "Social media certainly feeds oxygen to crackpot science."

7   As Schwartz and Thompson (1990: 33) argue:

> If different actors, in the same debate, cognize differently (that is, if they see things differently and know things differently), then they will inevitably be operating with different definitions of what is there. The debate, therefore, will entail the clash of differently drawn boundaries and the contention of incompatible rules of closure.

8   See Schwartz and Thompson (1990: 4–11), who cite Holling (1982) and Douglas (1982), and include an explanatory quadrant analysis.

9   See Mandel, Braman, and Kahan (2008).

10   Mandel, Braman, and Kahan (2008: 2).

11   Schwartz and Thompson (1990: 4).

12   Schwartz and Thompson (1990: 6; emphasis in original).

13   Schwarz and Thompson (1990: 18–19). The authors argue that "impact assessments, far from reflecting conflicting evaluations of the facts, involve rival *interpretive frames* in which facts and values are all bound up together" (Schwartz and Thompson, 1990: 23).

14   The term "crisis of expertise" was popularized by Gil Eyal (2019).

15   Other common examples of the politicization of science include the controversy over perceived dangers of COVID-19 vaccinations or the concern that the Trump administration downplayed or did not take the COVID-19 pandemic seriously.

16   Kofman (2018), quoting Bruno Latour in an interview with him.

17   See, for example, Groppe (2021).

18   See Stelter (2021).

19   See *Daubert v Merrell Dow Pharmaceuticals, Inc.* (1993) 590 U.S. 579.

20   Regarding the idealization of scientific expertise in the *Daubert* opinion and its aftermath, see, generally, Caudill and LaRue (2006).

21   See Caudill (2019).

22   See, generally, Eagleton (1991: 43–4). Marx and Engels viewed ideology negatively as the oppressive ideas of the ruling class that "give birth to a massive social illusion" (Eagleton, 1991: 43). Later Marxists like Lenin

would, however, speak of "socialist ideology," suggesting a broader use of the term—both the ruling class and the proletariat are ideological (see Eagleton, 1991: 44):

> Raymond Geuss has suggested a useful distinction between "descriptive", "pejorative" and "positive" definitions of the term ideology. In the descriptive or "anthropological" sense, ideologies are belief-systems characteristic of certain social groups or classes, composed of both discursive and non-discursive elements. We have seen already how this politically innocuous meaning of ideology comes close to the notion of a "world view", in the sense of a relatively well-systematized set of categories which provide a "frame" for the belief, perception and conduct of a body of individuals. (Eagleton, 1991: 43 [citing Guess, 1981: ch 1])

[23] See, for example, Smith (2018), arguing that the culture wars mirror the antagonism between pagans and Christians in the Roman Empire: As Lee (2021) writes:

> In the middle of the twentieth century, [as] sociologist Robert Wuthnow observed, where Christians had once distinguished themselves according to denominational identity, following the Second World War, institutional affiliation began to be overshadowed by political inclination. ...
>
> Fast-forward to the chaos of 2020, when political leanings have hardened into battle lines across American society, and the major Christian traditions have fragmented accordingly. Denominational identity is a forgotten relic, of interest primarily to veteran pastors and seminarians seeking ordination. The real question is whether you love Trump or despise him, whether you vote on abortion and religious liberty or racial justice and climate change.

[24] Castronuovo (2022). "The vague" guidelines were issued by The National Institutes of Health's COVID-19 Treatment Guidelines Panel, the Centers for Disease Control, and the Food and Drug Administration.

[25] Castronuovo (2022).

[26] Horst (2022: 461).

[27] Mihelj, Kondor, and Štětka (2022:293), citing Gustafson and Rice (2019). In more recent studies as well, "the perceived lack of expert consensus had a detrimental impact on perceived expertise" (Mihelj et al, 2022: 297).

[28] See Mihelj, Kondor, and Štětka (2022).

[29] See Mihelj, Kondor, and Štětka (2022).

[30] See Mihelj, Kondor, and Štětka (2022).

[31] See Mihelj, Kondor, and Štětka (2022).

[32] See Satta and Davidson (2019). For example, it is "the bad past actions of the medical profession, which in many places in the United States ... allowed and encouraged coerced sterilization of poor women of color, that is the source of ... distrust" (Satta and Davidson, 2019).

[33] Ceccarelli (2011: 196).

[34] Michaels (2008).

[35] These are Ceccarelli's examples of manufactured controversies (see Ceccarelli, 2011). The AIDS dissent in South Africa is also the primary example given by Weinel (2019).

[36] See, for example, Moreno and Holmgren (2014).

[37] See Ceccarelli (2011).

[38] See, generally, Weinel (2019).

[39] I am grateful to Dr Martin Weinel, Cardiff University, for sharing and discussing this schema with me.

[40] Email discussion with Dr Martin Weinel after my presentation (of the arguments in this book) at Cardiff University on February 21, 2021 (on file with the author).

[41] Gobo and Sena (2022: 25).

[42] Boulware et al (2022): "These errors, taken together, thoroughly discredit the ... claim that standard medical care for transgender children and adolescents constitutes child abuse."

[43] Boulware et al (2022): the drafters' findings "ignore established medical authorities and repeat discredited, outdated, and poor-quality information."

[44] Harambam (2020: 64), quoting Knight (2000: 95).

[45] Harambam (2020: 64), quoting Melley (2000: 20).

[46] See Harambam (2020: 5, 18).

[47] Harambam (2020: 18).

[48] Harambam (2020: 2).

[49] Pasquale (forthcoming).

[50] Pasquale (forthcoming).

[51] Pasquale (forthcoming).

[52] See, for example, Coglianese and Lai (2022: 1287): human "algorithms" undeniably fail due to memory limitations, cognitive biases, and groupthink.

[53] Pasquale (forthcoming).

[54] Abebe et al (2019).

[55] Pasquale (forthcoming).

[56] Abebe et al (2022).

one What Caused, and How Do We Fix, Our Crisis?

[1] See, for example, Tavernise (2020: A14). In response to a question by a reader whether she should stop speaking to Trump-supporting friends, the *New York Times* ethicist replied: "[P]eople can be epistemically disadvantaged by gaining their beliefs from social networks that are radically unreliable. We get many of our false beliefs … by listening to the views of people we trust" (Appiah, 2020: 20).

[2] Halbfinger (2020: A16). Representing the views of Nissim Mizrachi, Halbfinger (2020: A16) argues that both the working-class voters in Israel and the Trump voters "see themselves as their countries' most patriotic citizens, and demonize the left and its allies in the news media, academia and other liberal redoubts as traitorous enemies. Both … feel disdained by those elites, who dismiss their views as racist, ignorant or unwittingly self-defeating."

[3] Gilroy-Ware (2020: 63).

[4] Gilroy-Ware (2020: 209).

[5] Douthat (2020: SR9).

[6] See *The Scientist* (2020). With respect to COVID-19, a research article claiming that the virus was transmitted through surface contact (see Riddell et al, 2020) was found to be "a greatest-hits compilation of research errors" (Thompson, 2021) (grimy surfaces are not the problem; cleaning subways and buses every night is a waste of money).

[7] Thompson (2021),

[8] Latour (2018: 1).

[9] Latour, 2018: 2. Trump's withdrawal from the Paris Climate Accord "was a declaration of war authorizing the occupation of all the other countries, if not with troops, at least with CO2, which America retains the right to emit." Latour, 2018: 84. In doing so, Trump withdrew not only from the Paris Climate Accord, but from the earth, in order for the U.S. to occupy another imagined earth, and another imagined history in which "modernization" continues. Latour, 2018: 3–5. The actual earth, however, will react to that "action in such a way that [the U.S. will] no longer have a stable and indifferent framework in which to lodge [its] desires for modernization." Latour, 2018: 84.

[10] Latour (2018: 65).

[11] Latour (2018: 23).

[12] Latour (2018: 22, 25).

[13] Latour (2018: 25–6). To be clear, Latour certainly believes in consensus climate science, but only because he trusts processes taking place in socio-natural networks:

Latour was struck when he heard [a climatologist recently] defend his results not on the basis of the unimpeachable authority of science but by laying out … "the large number of researchers involved in climate analysis, the complex system for verifying data, the articles and reports, the principle of peer evaluation, the vast network of weather stations, floating weather buoys, satellites and computers that ensure the flow of information." The climate denialists, by contrast … had none of this institutional architecture. Latour realized he was witnessing [a] shift: from scientists appealing to transcendent, capital-T Truth to touting the robust networks through which truth is, and has always been, established. (Kofman, 2018)

[14] See *CNN Business* (2020).

[15] See: www.hbo.com/real-time-with-bill-maher/2020 (Accessed June 22, 2022).

[16] See Tavernise (2020: A14). When medical student Tho Nguyen used the word "brainwashed" to "describe her parents, her father said it applied to her … her parents did not believe Mr. Biden could have won, and it was hard to convince them otherwise, because that is not what they were hearing … on Facebook" (Tavernise, 2020: A14). Similarly, Danielle Ackley "got angry when she heard her mother criticize Mr. Biden's character. … 'This is not even a political divide, it's a reality divide,' said Ms. Ackley, who [saw] her mother comment approvingly on a Facebook post questioning mail-in ballots" (Tavernise, 2020: A14).

[17] See Mellon Seminar (2020).

[18] Eyal (2019: 4).

[19] Eyal (2019: 5–6).

[20] Eyal (2019: 143–4).

[21] Eyal (2019: 148–9). Regarding Breyer's proposal, see Breyer (1993).

[22] See *Daubert v Merrell Dow Pharmaceuticals, Inc.* (1993) 590 U.S. 579, creating a new regime for admissibility of expert testimony and establishing judges as gatekeepers in that regime.

[23] See Caudill and LaRue (2006).

[24] See Eyal (2019: 149): "The republic of trans-science would need to be one where 'bringing the bad news,' teaching others how to recognize 'inconvenient facts,' is established as a routine, yet honorable and well-regarded, vocation."

[25] See Latour (2018: 23).

[26] See Collins and Evans (2017: 4–9).

[27] See, generally, Collins et al (2020).

[28] See the opening remarks by Gil Eyal in Mellon Seminar (2020: 00:00–07:00).

[29] See the comments by Brubaker in Mellon Seminar (2020). The paradox is only apparent, however, because the indispensability of expertise and the high stakes of the pandemic make expertise vulnerable to attack—scientized politics brings with it politicized science. As to the latter point, see the comments by Brubaker (citing Eyal, 2019) in Mellon Seminar (2020).

[30] See the comments by Brubaker in Mellon Seminar (2020).

[31] See the comments by Brubaker in Mellon Seminar (2020). As Olena (2020) (quoting Patricia García, a Solidarity Trial investigator and the former health minister of Peru, speaking with *The Scientist* in the wake of the Surgisphere scandal [October 1, 2020 issue]) states:

> Now people are so confused about what science can give you—whether hydroxychloroquine works, it doesn't work, it's fake, it's not fake—that it's going to be very difficult for us scientists then to use any type of article or publication. Now that they know scientists can lie, who will believe us again?

[32] See the comments by Brubaker in Mellon Seminar (2020).

[33] Gilroy-Ware, 2020: 47.

[34] See the comments by Brubaker in Mellon Seminar (2020).

[35] See the comments by Hilgartner in Mellon Seminar (2020).

[36] See the comments by Hilgartner in Mellon Seminar (2020).

[37] See the comments by Hilgartner in Mellon Seminar (2020).

[38] See the comments by Hilgartner in Mellon Seminar (2020).

[39] See the comments by Hilgartner in Mellon Seminar (2020).

[40] See Durant (2019).

[41] Leonhardt (2022b). Public health officials "worry that people will misunderstand the details and behave dangerously" (Leonhardt, 2022b).

[42] See the comments by Tufekci in Mellon Seminar (2020).

[43] See the comments by Tufekci in Mellon Seminar (2020). See also Tufekci (2020): "interventions by authorities can backfire if they fuel mistrust or treat the public as an adversary rather than people who will step up if treated with respect."

[44] See the comments by Tufekci in Mellon Seminar (2020).

[45] See the comments by Lakoff in Mellon Seminar (2020). Brian Wynne (1989), for example, sees a need for citizen participation in scientific decision making, based on his famous study of Cumbrian sheep farmers whose sheep suffered due to fallout from the Chernobyl disaster. The sheep farmers knew more about how sheep walk around the pasture than the scientific experts studying the fallout and issuing a report, while the "experts" made mistakes because they did not know enough about the behaviors of sheep—hence Wynne's argument for citizen

participation. Cardiff sociologists of science Harry Collins and Robert Evans acknowledge the basis of Wynne's view in science and technology studies, which "provided a powerful argument against technocracy by showing how expert advice rested on a sea of social assumptions. This, in turn, led to arguments in favour of the democratization of science, and of expertise more generally" (Collins and Evans, 2017). Collins and Evans, however, reject the idea that the concerns of ordinary citizens should be reflected in science. In short, Wynne's sheep farmers were clearly *not* ordinary citizens, but rather experts in sheep farming—expertise does not require graduate degrees! Collins and Evans would limit the extension of technical decision-making rights to so-called "lay experts," an oxymoron in the context of public policy. We need experts (based on formal training *or* relevant experience) participating in scientific decision making, but citizens who are not experts in the relevant scientific field (in any public or regulatory debate) not only should be excluded, but also should be identified as illegitimate, external, political influences that should be resisted (Collins and Evans, 2017: 16–18).

46 See the comments by Hilgartner in Mellon Seminar (2020).

47 See the comments by Brubaker in Mellon Seminar (2020). This is similar to the concept of "epistemic humility": "Epistemic humility is an intellectual virtue. It is grounded in the realization that our knowledge is always provisional and incomplete—and that it might require revision in light of new evidence" (Angner, 2020).

48 Goldberg (2020).

49 Leonhardt (2022c).

50 Nilsen (2022).

51 Baumann (2022).

52 *Jacobson v Commonwealth* (1905), 197 U.S. 11.

53 *Jacobson v Commonwealth* (1905), 197 U.S. 11, p 26.

54 Burris (2021), discussing *Wisconsin Legislature v Palm* [2020] 942 N.W.2d 900., overturning Wisconsin's COVID-19 emergency measures, and *County of Butler v. Wolf* (2020) 486 F.Supp.3d 883, overturning the Pennsylvania governor's pandemic restrictions.

55 See *Klaasen v Trustees of Indiana University* (2021), 7 F.4th 592. The president of neighboring Purdue University said he would not institute a vaccine mandate, believing that it is up to each individual to decide whether to be vaccinated and, in any event, a mandate would be impractical and difficult to enforce (see MSNBC, 2021). Arizona Governor Doug Ducey signed an order on June 15, 2021, mandating that universities cannot require COVID-19 vaccinations, in which he also emphasized that despite his encouragement to get the vaccine, "it is a choice and we need to keep it that way." ABC.com (2021).

56 *Wall Street Journal* (2021).

57 See Senate Bill 2006 (2021 Florida Legislature), §18. Available at: www.flsenate.gov/Session/Bill/2021/2006/BillText/er/HTML (Accessed June 22, 2022).

58 See *Norwegian Cruise Line Holdings, Ltd. et al v Scott Rivkees, M.D.* (2021) 553 F.Supp 1143. The plaintiff is the holding company of Norwegian Cruise Line, Regent Seven Seas Cruises and Oceania Cruises; the defendant, Rivkees, is Florida's Surgeon General, who leads the Florida Department of Health.

59 First, what is considered a genuine religious belief in some states is not in others—veganism is not a religious belief in California, but it is in Ohio. See *Friedman v Southern California Permanente Medical Group* (2002) 102 Cal.App.4th 39 and *Chenzira v Cincinnati Children's Hospital Medical Center* [2012] Case No. 1:11-CV-00917, 2012 WL 6721098, U.S. Dist. Ct, S.D. Ohio, Western Division, December 27, in which the court found it "plausible that Plaintiff could subscribe to veganism with a sincerity equating that of traditional religious views." Second, while the "vast majority of states do not allow … philosophical exemptions," some do—Pennsylvania, for example, under the heading "Religious Exemption" (from immunization), allows objections based "on religious grounds or on the basis of strong moral or ethical conviction similar to religious belief" (see PA CODE § 23.84; see also Tomsick, 2020). As Tomsick (2020: 140) states:

> The states that do offer philosophical, or personal belief exemptions, employ a variety of procedures for parents to obtain the exemption. Exemption rates are significantly higher in the states where the exemption is more challenging to obtain. Some states require minimal effort—a parent may simply sign a form to exempt the child. In these states, exemption rates are high.

Finally, the experiences with immunizations of children indicate that "parents are using religious exemptions without really having a religious objection to vaccines. Religious exemptions are becoming a loophole" (Tomsick, 2020: 138).

60 See Hanson (2022).

61 Baker (2022).

62 See Cheng (2022).

63 Kahan, Jenkins-Smith, and Braman (20101: 2).

64 Kahan, Jenkins-Smith, and Braman (2011: 2).

65 Kahan, Jenkins-Smith, and Braman (2011: 3–4).

66 Kahan, Jenkins-Smith, and Braman (2011: 20–21).

67 Kahan, Jenkins-Smith, and Braman (2011:21, 24).

68 Kahan, Jenkins-Smith, and Braman (2011: 24).

69 Kahan, Jenkins-Smith, and Braman (2011: 23).
70 Kahan, Jenkins-Smith, and Braman (2011: 23).
71 Fischer (2021).
72 Fischer (2021: 1).
73 Fischer (2021: 2).
74 Gilroy-Ware (2020: 225).

two Worldviews as "Religious" Frameworks

1 Paine (1804).
2 I recognize, of course, that the opposing sides in the Protestant Reformation—Catholics and Protestants—do not capture the entire population of Europe or even Holland; there were also Jews and Muslims, and other religious and non-religious people, not involved in the Reformation as it swept across Europe.
3 Vanhaelen (2005: 251).
4 Emanuel de Witte, "Interior of the Oude Kerk, Delft," probably 1650, Metropolitan Museum of Art, color image available at: www.metmuseum.org/art/collection/search/438490 (the painting is reproduced on the cover of this book).
5 Calvinism stresses the isolation of each individual: "Each ... must travel [their] way of life alone. No preacher, no sacrament, no church can alter the inevitable destiny ordained of God" (Harkness, 1958: 182). The authority of the Church of Rome has here given way to individuals, who have direct interpretational access to the scriptures, the final authority (for the Reformers) on all issues. Witte (2008: 77) identifies in the Reformation a "fight for freedom" on the part of the individual against ecclesiastical powers.
6 In Abraham Kuyper's (1998 [1880]: 488) words: "There is not a square inch in the whole domain of our human existence over which Christ, who is Sovereign over all, does not cry: 'Mine!'"
7 That is the reason that "Dutch Calvinism did consider the possibility that a Christian merchant might not be a contradiction in terms" (Schama, 1987: 330). One does not enter the spiritual realm of church, prayer, and worship only to return to the "real" world of work and family, or even art—Christians can be "lovers of art and good Calvinists" (Vanhaelen, 2005: 259).
8 I am not arguing for disguised, moralizing messages that need to be deciphered—there is an open grave, likely symbolic, but that is not my focus. See Metropolitan Museum of Art (nd) ("a newly dug grave in the foreground provides a sobering reminder of mortality"). In my view, the description of the effects of Calvinist ideology is itself the "moralizing."

9   Vanhaelen (2005: 254, 258) notes that de Witte was "anything but an orthodox Christian," a warning to iconologists concerning biblical messaging.

10 De Jongh (2000:16), notwithstanding his influential iconological approach, concedes that "certain objects or motifs in seventeenth-century paintings often serve a dual function. They operate as concrete, observable things while at the same time doing something totally different, namely expressing an idea, a moral, an intention, a joke or a situation."

11 Vanhaelen (2005: 253). In his magnum opus, the four-volume *Institutes of the Christian Religion*, Calvin: (1) quoted the fourth-century Council of Elvira ("It is decreed that there shall be no pictures in churches, that what is reverenced or adored be not depicted on the walls"); (2) referred to Augustine's declaration that it is wrong to worship images; and (3) scolded the papists for their monstrous idols ("brothels show harlots clad more virtuously and modestly than the churches show these objects which they wish to be thought images of virgins") (Calvin, 2006 [1559]: Book 1, ch XI, §§ 5–6).

12 See Skillen (2014: 92). Hence, the "differences between Calvinism and Lutheranism can be accounted for in no small measure by the fact that Calvin began his career as a lawyer and Luther as a monk" (Harkness, 1958: 5).

13 Kuyper (1943: 7–8).

14 For example, Aquinas's definition of natural law, "which allows human reason a certain amount of autonomy in the moral realm, is absent from Calvin's work" (Backus, 2003: 12).

15 After studying and practicing law, Groen became active in politics—he was a member of the Second Chamber of Parliament for years (1849–57, 1862–66) (see Van Prinsterer, 1989 [1868]: 14; Schutte, 2005: 38).

16 Van Dijk (1975: vii).

17 See Kuyper (1943: 8).

18 As Bacote (2011: 24) writes: "Sphere sovereignty is Kuyper's idea that from God's sovereignty there derives more discrete sovereign 'spheres' such as the state, business, the family, and the church."

19 See Dooyeweerd (1969) and (1935–1936).

20 Zylstra (1975: 15–16).

21 Dooyeweerd (1948: v).

22 Dooyeweerd used the term "science" (*Wetenschap*) in the broad Continental sense of any knowledge and learning, including legal science.

23 Dooyeweerd (1948: v).

24 This view prefigured Polanyi's "framework of commitment" in which scientists work, Radnitsky's "steering fields" internal to science, and Kuhn's paradigm theory in the natural sciences (see Hart, 1985: 145, 150).

25 Latour (1993: 6).

26 See Latour (2018).
27 See de Vries (2016: 15).
28 Latour (2018: 24).
29 See Latour (2016): "Aesthetics" is "defined as what makes us sensitive to hitherto unknown phenomena."
30 Latour (2016). Latour refers elsewhere to the importance of novelists in 18th- and 19th-century "inventions" of democracy, class, and citizenship (see Latour, 2017).
31 Latour (2018: 26).
32 Kofman (2018).
33 See National Research Council Committee on Identifying the Needs of the Forensic Sciences Community (2009: 183–91).
34 See Kofman (2018):

> [O]ur current post-truth moment is less a product of Latour's ideas than a validation of them. In the way that a person notices her body only once something goes wrong with it, we are becoming conscious of the role that Latourian networks play in producing and sustaining knowledge only now that those networks are under assault.

See also Latour (2004: 226–7, 231):

> Do you see why I am worried? I myself have spent some time in the past trying to show "the lack of scientific certainty" inherent in the construction of facts. ... But I did not exactly aim at fooling the public by obscuring the certainty of a closed argument. ... I'd like to believe that, on the contrary, I intended to emancipate the public from prematurely naturalized objectified facts. ... The question was never to get away from facts but closer to them, not fighting empiricism but, on the contrary, renewing empiricism.

35 Shalizi (2017).
36 Taylor (2020: 317).

three The Quasi-Religious Aspect of the Crisis

1 Finkel et al (2020: 533).
2 See Finkel et al (2020)
3 Finkel et al (2020: 533).
4 See Finkel et al (2020: 533). See also Finkel et al (2020: 534):

The parties also have sorted along racial, religious, educational, and geographic lines. Although far from absolute, such alignment of ideological identities and demography transforms political orientation into a mega-identity that renders opposing partisans different from, even incomprehensible to, one another.

[5] Finkel et al (2020: 535).

[6] Finkel et al (2020: 535). As Finkel et al (2020: 535) writes: "[T]hree trends—identity alignment, the rise of partisan media, and elite ideological polarization—have contributed to" these different narratives. As to *identity alignment*: "alignment of ideological identities and demography transforms political orientation into a mega-identity that renders opposing partisans different from, even incomprehensible to, one another" (Finkel et al, 2020: 534). Regarding *elite ideological polarization*: "[I]n contrast to the equivocal ideological-polarization trends among the public, politicians and other political elites have unambiguously polarized recently on ideological grounds, with Republican politicians moving further to the right than Democratic politicians have moved to the left" (Finkel et al, 2020: 534).

[7] Finkel et al (2020: 534). People "who are already sectarian selectively seek out congenial news, but consuming such content also amplifies their sectarianism" (Finkel et al, 2020: 534).

[8] Finkel et al (2020: 536).

[9] Finkel et al (2020: 533).

[10] Hsu (2021: 405; emphasis added).

[11] Hsu (2021: 407).

[12] Hsu (2021: 410; emphasis in original). Hsu concedes that not all Republicans, and not all of those on the Right belonging to organized religions, have joined Trump in holding views hostile to science (see Hsu, 2021: 414).

[13] Hsu (2021: 411). Hsu (2021: 411) also states that "directing animus towards scientific experts and science is grotesquely misguided."

[14] Hsu (2021: 411–12).

[15] Hsu (2021: 412). As Hsu (2021: 413) states:

> The Green New Deal is a very broad and ambitious program created by the political left (some would say far left) to deal simultaneously with climate change and a variety of social and economic issues, and at times seems to be a basis for a Democratic Party litmus test. ... [I]ts proponents seem defiantly tone-deaf with respect to its fiscal implications, suggestive of resistance to or ignorance of economic science. [The idea that] "... the plan will pay for itself through economic growth" is similar to speculative

claims by the Trump Administration that federal government revenue lost by the 2017 Tax Cut and Jobs Act would be recaptured through economic growth.

[16] Hsu (2021: 443). Dr Anthony Fauci (quoted in Axelrod, 2021) recently discussed the Trump administration's attack on science:

> This is my seventh administration ... and I've been advising administrations and presidents on both sides of the aisle, Republicans and Democrats, people with different ideologies, and even with differences in ideology, there never was this real affront on science. So it really was an aberrancy that I haven't seen in almost 40 years that I've been doing this. So it's just one of those things that is chilling when you see it happen.

[17] Hsu (2021: 443). As Hsu (2021: 444) argues (citing Moffitt [2016], who states: "[P]opulists across the world have made headlines by setting 'the people' against the 'elite' in the name of popular sovereignty and 'defending democracy'"): "It is easy to portray scientists as part of a privileged 'elite,' a time-tested political epithet that has often been deployed to great effect in American political campaigns." See also Winberg (2017: 4).

[18] Hsu (2021: 448), citing Chua (2019: 137–64).

[19] Hsu (2021: 450–1).

[20] Sharfstein (2017).

[21] Sharfstein (2017).

[22] Sharfstein (2017), quoting Atul Gawande's graduation address at the California Institute of Technology in 2016.

[23] Sharfstein (2017).

four Belief as a Form of Expertise

[1] O'Connor (2020: 292).

[2] See Luhrmann (2020).

[3] Luhrmann (2020: 2). As Luhrmann (2020: 44) states: "Christians expect that prayer does not come easily and naturally. It is a skill that must be learned, as a relationship with God must also be learned."

[4] Luhrmann (2020: 17–18).

[5] Luhrmann (2020: 18), quoting Mair (2013: 449).

[6] Luhrmann (2020: 21). As Luhrmann (2020: 24) states: "People of faith live, in effect, on two levels ... attending to two different ways of making sense of the world."

[7]   Luhrmann (2020: 55, 57).

[8]   Luhrmann (2020: 58, 68, 70, 73, 84; emphasis in original). The ontological turn in anthropology recognized that "if we focus on belief, we tend to miss the experience" (Luhrmann, 2020: 183). There is faith, but there is also "the way the feeling of realness is kindled through practices, orientations, and the training of attention" (Luhrmann, 2020: 183).

[9]   Luhrmann (2020: 183).

[10] Wood (2020: 64).

[11] Wood (2020: 66).

[12] Wood (2020: 67).

[13] Wood (2020: 67).

[14] As Collins (2018: 68) states: "I can be an expert in astrology just as much as I can be an expert in astronomy."

[15] See Caudill et al (2019) and Collins and Evans (2002), suggesting that the future of STS would be to engage in "studies of expertise and experience."

[16] See Winch (1990).

[17] Collins (2018: 67), discussing Winch. Winch (1990: 42) explains Wittgenstein's concept of *form of life* and its relation to sociology as follows:

> [T]o understand the nature of social phenomena in general, to elucidate ... the concept of a "form of life," has been shown to be precisely the aim of epistemology. [T]he epistemologist's starting point is rather different from that of the sociologist but, if Wittgenstein's arguments are sound, that is what he must sooner or later concern himself with. That means the relations between sociology and epistemology must be ... very much closer than ... what is usually imagined to be the case.

Here, Winch is disagreeing with analytic philosophers A.J. Ayer and P.F. Strawson, whose reading of Wittgenstein does not fully appreciate that a "single use of language does not stand alone; it is intelligible only within the general context in which language is used" (Winch, 1990: 39). For another account of the importance of Wittgenstein for social studies of science, see Bloor (1983), proposing that Wittgenstein's later philosophy should be interpreted as a social theory of knowledge.

[18] Collins (2018: 68).

[19] Collins (2018: 68).

[20] Collins (2018: 68). See also Collins (2019: 147):

> [A]n expertise was something you were good at ... but the thing you were good at could be anything, as long as members of the domain believed it was worth being good at it: it didn't have to

be anything true or have useful consequences. What you were good at could be reading tea leaves or econometric modeling of economics and it was still an expertise.

21 Eyal (2019: 131), quoting Edwards (2010: 407). As Eyal (2019: 135) states: "When activists complained about 'industry bias' [at the FDA], they reinforced the FDA's image as protector of the public. When industry complained of 'over caution,' it reinforced the perception of the agency as composed of careful medical specialists."

22 Eyal (2019: 136–7). One strategy of industries trying to avoid governmental regulation based on the potential harm to the public of their operations is to claim that the scientific evidence is not sufficient; this gave rise to the "sound science" movement on the political Right, demanding *more* science in order justify new regulations.

23 Wittgenstein (1958: §§ 19, 23; emphasis in original). The term "forms of life":

> is evidently meant to refer to the stories that are behind the words, and which illuminate the rules by which the words acquire their meaning. ... The forms of life are ... the manner of action shared by people of a particular time and culture. Wittgenstein ... firmly rejects the idea that it is a question merely of what people agree on. He distinguishes here between agreement of opinions and agreement in form of life. It becomes clear ... that he is really concerned with fundamental attitudes ... [particularly] to our attitude towards other people. (Van Peursen, 1969: 108–9)

Human beings agreeing on the language they use is not, Wittgenstein (1958: § 241) says, "agreements in opinion, but in form of life."

24 Hacker (2015: 1–4). As to the controversy over the implications of Wittgenstein's use of the term "*Lebensform*," it bears mention that interpretations of Wittgenstein's work vary. Offering the varied interpretations of Hegel as an example, Bloor (1992: 281) writes: "The time is clearly past when it was useful to speak of a position being simple 'Wittgensteinian' or 'non-Wittgensteinian.' There are different and opposed readings of Wittgenstein, and different and opposed lessons drawn from his work. Such a situation is not surprising, and there are well-known precedents."

25 Hunter (1968: 233–4). As Moyal-Sharrock (2016: 37–9) writes:

> [T]he view that forms of life are synonymous, or quasi-synonymous, with language or language-games ... is due, I believe, to a misunderstanding of Wittgenstein's remark that "to imagine

a language means to imagine a form of life" [Wittgenstein, 1958: §19]. The view is discredited by most commentators, but not all ...

Further:

> When Wittgenstein writes that "to imagine a language means to imagine a form of life," he does not mean to equate both, but to suggest that language is logically connected to a form of life: there can be no language without a form of life from which it can spring, and which provides the necessary context for expressing meaning. (Moyal–Sharrock, 2016: 37)

For example, being religious "is a form of life, but speaking religiously is not; speaking religiously is speaking from the perspective of a religious form of life" (Moyal–Sharrock, 2016: 38).

[26] Hunter (1968: 234).
[27] Hunter (1968: 234).
[28] Hunter (1968: 235).
[29] Moyal–Sharrock (2016: 28–32).
[30] Moyal–Sharrock (2016: 28–32), discussing (1) the organic or biological account of Garver (1994) and (2) the historico–cultural view of Baker and Hacker (2009a, 2009b).
[31] Baker and Hacker (2009a: 74). A form of life "includes shared natural and linguistic responses, broad agreement in definitions and judgements, and corresponding behaviour" (Baker and Hacker, 2009a: 74).
[32] Moyal–Sharrock (2016: 27).
[33] As Cavell (1969: 172) writes: "The religious ... should be thought of as a Wittgensteinian form of life."
[34] Cavell (1969: 172).
[35] Moyal–Sharrock (2016: 38; emphasis added).

five Communicating across Worldviews

[1] Nunez (2020: 183).
[2] Collins (2020: 933).
[3] Toner (2017: 15).
[4] Toner (2017: 16–17).
[5] Wittgenstein (1969: 34e).
[6] Toner (2017: 17).
[7] Toner (2017: 17).
[8] See Malcolm (1976: 72), opining that Wittgenstein saw religion as a form of life.

9 Keightley (1976: 49–50), citing Wittgenstein (1971: 29–30).

10 Keightley (1976: 50–1).

11 Keightley (1976: 53). As Keightley (1976: 52) writes: "The kind of disagreement between believer and unbeliever is so fundamental that their disagreement cannot be located within any mode of discourse."

12 Wittgenstein (1966: 55).

13 Keightley (1976: 52).

14 Keightley (1976: 71–2).

15 Keightley (1976: 102–3). As Winch (1990: 79–81) writes:

> [T]o understand the activities of an individual scientific investigator we must take into account two sets of relations: first, his relation to the phenomenon which he investigates; second, his relation to his fellow-scientists. Both of these are essential to the sense of saying that he is … "discovering uniformities"; but writers on scientific "methodology" too often concentrate on the first and overlook the importance of the second. … [When the scientist] on the basis of his observations of the phenomenon (in the course of his experiments) … develops his concepts … he is able to do this only in virtue of his participation in an established form of activity with his fellow-scientists. … [T]hey are all taking part in the same general kind of activity, which they have all *learned* in similar ways … they are, therefore, *capable* of communicating with each other about what they are doing.

16 Winch (1970: 53; emphasis in original).

17 Keightley (1976: 104), discussing Kuhn (1973).

18 Collins (2016: 68).

19 As Collins (2016: 72) argues, someone who does not practice in a domain "but acquires only the spoken discourse of the domain … is an interactional expert."

20 Collins embedded himself in the gravitational wave community and became an interactional expert in the field (see Collins, 2017). As Collins (2017: 371) writes: "The idea of interactional expertise brings language into central focus. … [B]y spending enough time taking part in the spoken discourse of a specialist group—by acquiring their practice language … one can learn to understand their world of practice without taking part in the practice itself." Collins and others have explored the notion of interactional expertise in imitation games based on the Turing test; it was in one of those games that his knowledge of gravitational wave physics was tested:

We use imitation games, that is, Turing tests with humans, to test for the possession of interactional expertise. ... The original gravitational wave imitation game test was run over email, with a gravitational wave physicist setting seven questions that were answered by me along with another gravitational wave physicist. The completed dialogues were sent to nine other gravitational wave physicists who were asked: "Which is the real gravitational wave physicist and which is Harry Collins?" Seven of these said they could not tell the difference, and two said Collins was the real [one]. (Collins, 2017: 374)

[21] Leonhardt (2022a).

[22] See Davies (2021: 116–33).

[23] National Research Council Committee on the Science of Science, Division of Behavioral and Social Sciences and Education (2017: 52).

[24] Horst (2022: 461).

[25] As Hochschild (2016: xii) writes: "I felt like I was in a foreign country, only this time it was my own."

[26] See Hochschild (2016: 166–7, 176–7, 183–4).

[27] Hochschild (2016: 6).

[28] Hochschild (2016: 7).

[29] Hochschild (2016: xi–xii).

[30] National Research Council Committee on the Science of Science, Division of Behavioral and Social Sciences and Education (2017: 56).

[31] National Research Council Committee on the Science of Science, Division of Behavioral and Social Sciences and Education (2017: 56).

[32] As Douglas (2005: 15) writes: "a clear demarcation between the realm of science and the realm of policy as understood in this way is not desirable (even if it were achievable)."

[33] Collins and Evans (2017: 124).

[34] Kofman (2018), quoting Latour. The traditional conception of nature cannot be politicized, Latour observes, since it was "invented precisely to limit human action thanks to an appeal to the laws of objective nature that cannot be questioned" (Latour, 2018: 65).

[35] Kofman (2018), paraphrasing Latour.

[36] National Research Council Committee on the Science of Science, Division of Behavioral and Social Sciences and Education (2017: 56).

[37] National Research Council on the Science of Science, Division of Behavioral and Social Sciences and Education (2017: 62).

[38] National Research Council Committee on the Science of Science, Division of Behavioral and Social Sciences and Education (2017: 64), citing Lewandowsky et al (2012).

39 National Research Council Committee on the Use of Social Science Knowledge in Public Policy, Division of Behavioral and Social Sciences and Education (2012: 226).
40 Fischer and Gottweis (2012: 1).
41 Fischer and Gottweis (2012: 1–2).
42 Fischer and Gottweis (2012: 9).
43 Fischer and Gottweis (2012: 10).
44 Fischer and Gottweis (2012: 14).
45 Fischer (2013: 100).
46 Fischer and Gottweis (2012: 6). I am, that is, slower to reject "simplified academic models of explanation" (Fischer and Gottweis, 2012: 6) if those models include consensus science.
47 Oreskes (2019: 4).
48 See Oreskes (2019: 142, 247).
49 See Richards (2010).
50 See Oreskes (2019: 144).
51 Oreskes (2019: 137).

Conclusion

1 Foster (2012), citing Latour (2004).
2 Latour (2004: 241).
3 Collins (2017: 376).
4 Reverend Stephen W. Planning, SJ, "Message from President of Gonzaga College High School, Washington D.C., January 7, 2021" (on file with author), continuing: "When we encounter a person who thinks differently than us, St. Ignatius would encourage us to choose to believe that a person who, on the surface seems so different from me, usually desires many of the same things I do."
5 Collins (2019: 65, 70).
6 Indeed, the so-called "Science Wars" between scholars in STS and those with traditional, idealistic images of science appeared to be "a prelude to the post-truth era," culminating in the "precipitous rise ... in anti-scientific thinking," the "death of expertise," and the election of a US "president who invents the facts to suit his mood and goes after the credibility of anyone who contradicts him" (Kofman, 2018). Steve Fuller (2016) even accepts responsibility for a post-truth world as the legacy of STS:

> My own view has always been that a post-truth world is the inevitable outcome of greater epistemic democracy. In other words, once the instruments of knowledge production are made generally available—and they have been shown to work—they

will end up working for anyone with access to them. ... [W]e should finally embrace our responsibility for the post-truth world.

7   See Caudill and LaRue (2003).
8   Kristof (2021: A13).
9   Kristof (2021: A13).
10 Spivak (1990: 41).

References

ABC.com (2021) "Arizona executive order bans universities from requiring students to receive COVID-19 vaccine," June 15. Available at: www.abc15.com/news/state/gov-ducey-issues-order-for-college-students-to-not-be-mandated-to-take-covid-19-vaccine (Accessed June 22, 2022).

Abebe, R., Barocas, S., Kleinberg, J., Levy, K., Raghavan, M., and Robinson, D.G. (2019) "Roles for computing in social change," paper presented at the ACM Conference on Fairness, Accountability, and Transparency, May 12. Available at: https://ssrn.com/abstract=3503029 (Accessed June 22, 2022).

Abebe, R., Hardt, M., Jin, A., Miller, J., Schmidt, L., and Wexler, R. (2022) "Adversarial scrutiny of evidentiary statistical software," in *Proceedings of the Conference on Fairness, Accountability, and Transparency*. Available at: https://ssrn.com/abstract=4107017 (Accessed June 22, 2022).

Angner, E. (2020) "Epistemic humility—knowing your limits in a pandemic," *Behavioral Scientist*, April 13. Available at: https://behavioralscientist.org/epistemic-humility-coronavirus-knowing-your-limits-in-a-pandemic/ (Accessed June 22, 2022).

Appiah, K.A. (2020) "Should I stop speaking to my Trump-supporting friends?," *The New York Times Magazine*, December 6, p 20.

Axelrod, T. (2021) "Fauci describes 'chilling' pressure on scientists in Trump era," *The Hill*, January 23. Available at: https://thehill.com/policy/healthcare/535515-fauci-describes-chilling-pressure-on-scientists-in-trump-era/ (Accessed June 22, 2022).

Backus I. (2003) "Calvin's concept of natural and Roman law," *Calvin Theological Journal*, 38: 7–26.

Bacote, V.E. (2011) "Introduction," in J.J. Ballor and S.J. Grabill (eds) *Wisdom and Wonder: Common Grace in Science and Art* (trans N.D. Klossterman), Grand Rapids, MI: Christian's Library Press, pp 23–9.

Baker, G.P. and Hacker, P.M.S. (2009a) *Wittgenstein: Rules, Grammar and Necessity: Essays and Exegesis of §§ 1–184, Volume 1*, 2nd edn, Oxford: Wiley Blackwell.

Baker, G.P. and Hacker, P.M.S. (2009b) *Wittgenstein: Rules, Grammar and Necessity: Essays and Exegesis of §§ 185–242, Volume 2*, 2nd edn, Oxford: Wiley Blackwell.

Baker, J. (2022) "Cincinnati federal judge blocks Air Force, Air National Guard globally from discharging religious vaccine refusers," *WXIX Cincinnati*, July 28. Available at: www.msn.com/en-us/news/us/cincinnati-federal-judge-blocks-air-force-air-national-guard-globally-from-discharging-religious-vaccine-refus ers/ar-AA102I4L (Accessed October 26, 2022).

Baumann, J. (2022) "Mask ruling appeal gets backing from former CDC directors," *Bloomberg Law*, June 7. Available at: https://news.bloomberglaw.com/health-law-and-business/mask-ruling-app eal-gets-backing-from-former-cdc-directors-16 (Accessed June 22, 2022).

Bloor, D. (1983) *Wittgenstein: A Social Theory of Knowledge*, New York: Columbia University Press.

Bloor, D. (1992) "Left and right Wittgensteinians," in A. Pickering (ed) *Science as Practice and Culture*, Chicago, IL: University of Chicago Press, pp 266–82.

Boulware, S. Kamody, R., Kuper, L., McNamara, M., Olezeski, C., Szilagyi, N. and Alstott, A. (2022) "Biased science: the Texas and Alabama measures criminalizing medical treatment for transgender children and adolescents rely on inaccurate and misleading scientific claims," Yale Law School Public Law Research Paper, April 28. Available at: https://papers.ssrn.com/sol3/papers.cfm?abstract_id=4102374 (Accessed October 26, 2022).

Breyer, S. (1993) *Breaking the Vicious Circle: Toward Effective Risk Regulation*, Cambridge, MA: Harvard University Press.

Burris, S. (2021) "Individual liberty, public health, and the battle for the nation's soul," *The Regulatory Review*, June 7. Available at: www.theregreview.org/2021/06/07/burris-individual-liberty-public-health-battle-for-nations-soul/ (Accessed June 22, 2022).

Calvin, J. (2006 [1559]) *Institutes of the Christian Religion* (ed J.T. McNeill; trans F.L. Battles), Louisville, KY: Westminster John Knox Press.

Castronuovo, C. (2022) "Pfizer Covid pill access stymied by 'vague' prescribing guidance," *Bloomberg Law (Deep Dive)*, June 3. Available at: https://news.bloomberglaw.com/health-law-and-business/pfizer-covid-pill-access-stymied-by-vague-prescribing-guidance (Accessed June 22, 2022).

Caudill, D.S. (2019) "Twenty-five years of opposing trends: the demystification of science in law, and the waning relativism in the sociology of science," in D.S. Caudill, S.N. Conley, M.E. Gorman, and M. Weinel (eds) *The Third Wave of Science and Technology Studies: Future Research Directions on Expertise and Experience*, New York: Palgrave Macmillan, pp 17–32.

Caudill, D.S. and LaRue, L.H. (2003) "Why judges applying the Daubert trilogy need to know about the social, institutional, and rhetorical, and not just the methodological, aspects of science," *Boston College Law Review*, 45: 1–53.

Caudill, D.S., Conley, S.N., Gorman, M.E. and Weinel, M. (2019) "Introduction," in D.S. Caudill, S.N. Conley, M.E. Gorman, and M. Weinel (eds) *The Third Wave of Science and Technology Studies: Future Research Directions on Expertise and Experience*, New York: Palgrave Macmillan, pp 1–16.

Cavell, S. (1969) *Must We Mean What We Say? A Book of Essays*, Cambridge: Cambridge University Press.

Ceccarelli, L. (2011) "Manufactured scientific controversy: science, rhetoric, and public debate," *Rhetoric and Public Affairs*, 14(2): 195–228.

Cheng, E.K. (2022) "The consensus rule: a new approach to scientific evidence," *Vanderbilt Law Review*, 75: 407–74.

Chua, A. (2019) *Political Tribes: Group Instinct and the Fate of Nations*, London: Penguin Press.

CNN Business (2020) "Trump's election-denying tweets are part of an ecosystem," November 22. Available at: www.cnn.com/videos/business/2020/11/22/trumps-election-denying-tweets-are-part-of-an-ecosystem.cnn (Accessed June 22, 2022).

Coglianese, C. and Lai, A. (2022) "Algorithm vs. algorithm," *Duke Law Journal*, 72: 1281–340.

Collins, H. (2017) *Gravity's Kiss: The Detection of Gravitation Waves*, Cambridge, MA: MIT Press.

Collins, H. (2018) "Studies of expertise and experience," *Topoi*, 37(1): 67–77.

Collins, H. (2019) *Forms of Life: The Method and Meaning of Sociology*, Cambridge, MA: MIT Press.

Collins, H. (2020) "Interactional Imogen: language, practice and the body," *Phenomenology and Cognitive Science*, 19: 933–60. Available at: https://doi.org/10.1007/s11097-020-09679-x (Accessed June 22, 2022).

Collins, H. and Evans, R. (2002) "The Third Wave in science studies," *Social Studies of Science*, 32(2): 235–96.

Collins, H. and Evans, R. (2017) *Why Democracies Need Science*, Cambridge: Polity Press.

Collins, H., Evans, R., Durant, D., and Weinel, M. (2020) *Experts and the Will of the People: Society, Populism and Science*. New York: Palgrave Pivot.

Cook, J. (2016) "Countering climate science denial and communicating scientific consensus," *Oxford Research Encyclopedias of Climate Science*. Available at: https://doi.org/10.1093/acrefore/9780190228620.013.314 (Accessed June 22, 2022).

Davies, S.R. (2021) "An empirical and conceptual note on science communication's role in society," *Science Communication*, 43(1): 116–33.

De Jongh, E. (2000) *Questions of Meaning: Theme and Motif in Dutch Seventeenth-Century Painting* (ed M. Hoyle; trans M. Hoyle), Leiden: Primavera Press.

De Vries, G. (2016) *Bruno Latour*, Cambridge: Polity Press.

Dooyeweerd, H. (1935–1936) *De wijsbegeerte der wetsidee* (3 vols), Amsterdam: H.J. Paris.

Dooyeweerd, H. (1948) *Transcendental Problems of Philosophic Thought*, Grand Rapids, MI: Eerdmans.

Dooyeweerd, H. (1969) *A New Critique of Theoretical Thought* (4 vols), Philadelphia, PA: Presbyterian & Reformed Pub.

Douglas, H. (2005) "Boundaries between science and policy: descriptive difficulty and normative desirability," *Environmental Philosophy*, 2(1): 14–29.

Douglas, M. (1982) *Essays in the Sociology of Perception*, New York: Routledge.

Douthat, R. (2020) "When you can't just 'trust the science'," *The New York Times*, December 20, p SR9.

Durant, D. (2019) "Ignoring expertise," in D.S. Caudill, S.N. Conley, M.E. Gorman, and M. Weinel (eds) *The Third Wave in Science and Technology Studies: Future Research Directions on Expertise and Experience*, New York: Palgrave Macmillan, pp 33–52.

Eagleton, T. (1991) *Ideology: An Introduction*, London: Verso.

Edwards, P.N. (2010) *A Vast Machine: Computer Models, Climate Data, and the Politics of Global Warming*, Cambridge, MA: MIT Press.

Eyal, G. (2019) *The Criss of Expertise*, Cambridge: Polity Press.

Finkel, E.J., Bail, C.A., Cikara, M., Ditto, P.H., Iyengar, S. Klar, S., Mason, L., McGrath, M.C., Nyhan, B., Rand, D.G., Skitka, L.J., Tucker, J.A., Van Bavel1, J.J., Wang, C.S. and Druckman, J.N. (2020) "Political sectarianism in America: a poisonous cocktail of othering, aversion, and moralization poses a threat to democracy," *Science*, 370(6516): 533–6. Available at: www.science.org/doi/abs/10.1126/science.abe1715 (Accessed June 22, 2022).

Fischer, F. (2013) "Policy expertise and the argumentative turn: toward a deliberative policy-analytic approach," *Revue française de science politique (English Edition)*, 63(3–4): 95–114.

Fischer, F. (2021) *Truth and Post-truth in Public Policy*, Cambridge: Cambridge University Press.

Fischer, F. and Gottweis, H. (2012) "Introduction," in F. Fischer and H. Gottweis (eds) *The Argumentative Turn Revisited: Public Policy as Communicative Practice*, Durham, NC: Duke University Press, pp 1–27.

Foster, H. (2012) "Post-critical," *The Brooklyn Rail: Critical Perspective on Arts, Politics, and Culture*, December 12–January 13. Available at: https://brooklynrail.org/2012/12/artseen/post-critical (Accessed June 22, 2022).

Fuller, S. (2016) "Embrace the inner fox: post-truth as the STS symmetry principle universalized," *Social Epistemology Review and Reply Collective*, December 25. Available at: https://social-epistemology.com/2016/12/25/embrace-the-inner-fox-post-truth-as-the-sts-symmetry-principle-universalized-steve-fuller/ (Accessed June 22, 2022).

Garver, N. (1994) *This Complicated Form of Life: Essays on Wittgenstein*, Chicago, IL: Open Court.

Gilroy-Ware, M. (2020) *The Truth about Fake News*, London: Repeater Books.

Gobo, G. and Sena, B. (2022) "Questioning and disputing vaccination policies. Scientists and experts in the Italian public debate," *Bulletin of Science, Technology & Society*, 42(1–2): 25–38.

Goldberg, J. (2020) "Would the Republican Party consider throwing Trump overboard?," *Tribune Content Agency*, December 28. Available at: www.reviewjournal.com/opinion/jonah-goldberg-would-the-republican-party-consider-throwing-trump-overbo ard-2069015/ (Accessed October 26, 2022).

Groppe, M. (2021) "Fauci unleashed: he says it's 'liberating' that he can 'let the science speak' as adviser to Biden," *USA TODAY*, January 22. Available at: www.usatoday.com/story/news/polit ics/2021/01/21/anthony-fauci-speaking-covid-liberating-under-biden-vs-trump/4244169001/ (Accessed October 26, 2022).

Guess, R. (1981) *The Idea of Critical Theory: Habermas and the Frankfurt School*, Cambridge: Cambridge University Press.

Gustafson, A. and Rice, R.E. (2019) "The effects of uncertainty frames in three science communication topics," *Science Communication*, 41(6): 679–706.

Hacker, P.M.S. (2015) "Forms of life," *Nordic Wittgenstein Review*, Special Issue: 1–4.

Hadhazy, A. (2010) "Anti-vaccination groups dealt blow as *Lancet* study is retracted," *Popular Mechanics*, February 5. Available at: www.popularmechanics.com/science/health/a5008/4344963/ (Accessed June 22, 2022).

Halbfinger, D.M. (2020) "Explaining right-wing politics in America, via the Middle East," *The New York Times*, December 19, p A16.

Hanson, E. (2022) "Judge orders parents to pay hefty legal fees in dismissed school mask case," *Go2Tutors*, July 28. Available at: www.msn.com/en-us/news/crime/judge-orders-parents-to-pay-hefty-legal-fees-in-dismissed-school-mask-case/ar-AA104FbY (Accessed October 26, 2022).

Harambam, J. (2020) *Contemporary Conspiracy Culture: Truth and Knowledge in an Era of Epistemic Instability*, Abingdon: Routledge.

Harkness, G. (1958) *John Calvin: The Man and His Ethics*, Nashville, TN: Abingdon Press.

Hart, H. (1985) "Dooyeweerd's Gegenstand theory of theory," in C.T. McIntire (ed) *The Legacy of Herman Dooyeweerd: Reflections on Critical Philosophy in the Christian Tradition*, Lanham, MD: University Press of America, pp 143–66.

Hochschild, A.R. (2016) *Strangers in Their Own Land: Anger and Mourning on the American Right*, New York: The New Press.

Holling, C.S. (1982) "Myths of ecological stability," in G. Smart and W. Stansbury (eds) *Studies in Crisis Management*, Montreal: Butterworth, pp 93–106.

Horst, M. (2022) "Science communication as a boundary space: an interactive installation about the social responsibility of science," *Science, Technology, & Human Values*, 47(3): 459–82.

Hsu, S. (2021) "Anti-science ideology," *University of Miami Law Review*, 75: 405–58.

Hunter, J.F.M. (1968) " 'Forms of life' in Wittgenstein's philosophical investigations," *American Philosophical Quarterly*, 5(4): 233–43.

Kahan, D.M., Jenkins-Smith, H., and Braman, D. (2011) "Cultural cognition of scientific consensus," *Journal of Risk Research*, 4(2): 147–74.

Keightley, A. (1976) *Wittgenstein, Grammar and God*, London: Epworth Press.

Knight, P. (2000) *Conspiracy Culture: From the Kennedy Assassination to the X-Files*, Abingdon: Routledge.

Kofman, A. (2018) "Bruno Latour, the post-truth philosopher, mounts a defense of science," *The New York Times Magazine*, October 25. Available at: www.nytimes.com/2018/10/25/magaz ine/bruno-latour-post-truth-philosopher-science.html (Accessed June 22, 2022).

Kristof, N. (2021) "Defeating racism one conversation at a time," *The New York Times*, July 10, p A13.

Kuhn, T. (1973) *The Structure of Scientific Revolutions*, Chicago, IL: University of Chicago Press.

Kuyper, A. (1943) *Lectures on Calvinism*, Grand Rapids, MI: Eerdmans.

Kuyper, A. (1998 [1880]) "Sphere sovereignty (inaugural address at the dedication of the Free University Amsterdam, 1880)," in J.D. Bratt (ed) *Abraham Kuyper: A Centennial Reader*, Grand Rapids, MI: Eerdmans, pp 461–90.

Latour, B. (1993) *We Have Bever Been Modern* (trans C. Porter), Cambridge, MA: Harvard University Press.

Latour, B. (2004) "Why has critique run out of steam? From matters of fact to matters of concern," *Critical Inquiry*, 30(4): 225–48.

Latour, B. (2016) "On sensitivity: arts, science and politics in the new climatic regime," keynote lecture at the University of Melbourne for the opening of the Performance Studies International, July 5. Available at: www.bruno-latour.fr/node/692 (Accessed June 17, 2020).

Latour, B. (2017) "What are the optimal interrelations of art, science and politics in the Anthropocene?," *Bifrost Insights*, November 30. Available at: https://bifrostonline.org/bruno-latour-what-are-the-optimal-interrelations-of-art-science-and-politics-in-the-anthropocene/ (Accessed June 17, 2020).

Latour, B. (2018) *Down to Earth: Politics in the New Climatic Regime* (trans C. Porter), Cambridge: Polity Press.

Lee, G. (2021) "The complexities of forbearance: Augustinian insights for an age of polarization," *Comment*, January 14. Available at: www.cardus.ca/comment/article/the-complexities-of-forb earance/ (Accessed June 22, 2022).

Leonhardt, D. (2022a) "What Benjamin Franklin's vaccine experience tells us about convincing the unconvinced," *The New York Times*, March 3. Available at: www.nytimes.com/2022/03/03/briefing/covid-vaccination-ben-franklin-ken-burns.html (Accessed June 22, 2022).

Leonhardt, D. (2022b) "The federal government is telling us two different stories about Covid vaccines for young children," *The New York Times*, April 29. Available at: www.nytimes.com/2022/04/29/briefing/vaccines-kids-moderna-pfizer.html (Accessed June 22, 2022).

Leonhardt, D. (2022c) "Masks work. So why haven't Covid mask mandates made much difference?," *The New York Times*, May 31. Available at: www.nytimes.com/2022/05/31/briefing/masks-mandates-us-covid.html (Accessed June 22, 2022).

Lewandowsky, S., Ecker, U.K.H., Seifert, C., Schwarz, N. and Cook, J. (2012) "Misinformation and its correction continued influence and successful debiasing," *Psychological Science in the Public Interest*, 13(3): 106–31.

Luhrmann, T.M. (2020) *How God Becomes Real: Kindling the Presence of Invisible Others*, Princeton, NJ: Princeton University Press.

Luntz, F. (2002) *The Environment: A Cleaner, Safer, Healthier America*, Alexandria, VA: Luntz Research.

Mair, J. (2013) "Cultures of belief," *Anthropological Theory*, 12(4): 448–66.

Malcom, N. (1976) *Ludwig Wittgenstein: A Memoir*, London: Oxford University Press.

Mandel, G., Braman, D., and Kahan, D. (2008) "The cultural condition of synthetic biology risks: a preliminary analysis," *Cultural Cognition Project at Yale Law School*. Available at: http://ssrn.com/abstract=1264804 (Accessed June 22, 2022).

Mastrangelo, D. (2020) "Gingrich won't accept Biden as president, says Democrats, Republicans 'live in alternative worlds'," *The Hill*, December 23. Available at: https://thehill.com/homenews/531261-gingrich-wont-accept-biden-as-president-says-democrats-and-republicans-live-in/ (Accessed June 22, 2022).

McLaughlin, K. and Dzhanova, Y. (2020) "Experts warn anti-vaxxer concerns about a COVID-19 vaccine could slow the end of the pandemic," *Business Insider*, December 3. Available at: www.busi nessinsider.nl/experts-warn-anti-vaxxer-concerns-about-a-covid-19-vaccine-could-slow-the-end-of-the-pandemic/ (Accessed October 26, 2022).

Melley, T. (2000) *Empire of Conspiracy: The Culture of Paranoia in Postwar America*, Ithaca, NY: Cornell University Press.

Mellon Seminar (2020) "Experts, publics and trust during the pandemic: sociological perspectives," October 29. Available at: www.dropbox.com/sh/gfsut0sbu2psmnb/AAA1V9fEP5R_ G2yIjV04BcXNa?dl=0&preview=GMT20201029-160134_Expe rts--P_1920x1080.mp4

Metropolitan Museum of Art (nd) "Browse the collection." The New York Metropolitan Museum of Art's iconographic viewer's guide to Interior of the Oude Kerk, Delft. Available at: www. metmuseum.org/art/collection/search/438490 (Accessed June 22, 2022).

Michaels, D. (2008) *Doubt Is Their Product: How Industry's Assault on Science Threatens Your Health*, Oxford: Oxford University Press.

Mihelj, S., Kondor, K., and Štětka, V. (2022) "Establishing trust in experts during a crisis: expert trustworthiness and media use during the COVID-19 pandemic," *Science Communication*, 44(3): 292–319.

Moffitt, B. (2016) *The Global Rise of Populism: Performance, Political Style, and Representation*, Stanford, CA: Stanford University Press.

Moreno, J.A. and Holmgren, B. (2014) "The Supreme Court screws up the science: there is no abusive head trauma/shaken baby syndrome 'scientific' controversy," *Utah Law Review*, 2013(5): 1357–453.

Moyal-Sharrock, D. (2016) "Wittgenstein on forms of life, patterns of life and ways of living," *Nordic Wittgenstein Review*, Special Issue: 21–42.

MSNBC (2021) "Purdue University president reacts to judge's ruling to uphold nearby Indiana University's vaccine mandate," July 19. Available at: www.msnbc.com/stephanie-ruhle/watch/purdue-university-president-reacts-to-judge-s-ruling-to-uph old-nearby-indiana-university-s-vaccine-mandate-117007429 838 (Accessed October 26, 2022).

National Research Council Committee on Identifying the Needs of the Forensic Sciences Community (2009) *Strengthening Forensic Science in the United States: A Path Forward*, Washington, DC: The National Academies Press.

National Research Council Committee on the Science of Science, Division of Behavioral and Social Sciences and Education (2017) *Communicating Science Effectively: A Research Agenda*, Washington, DC: National Academies Press.

National Research Council Committee on the Use of Social Science Knowledge in Public Policy, Division of Behavioral and Social Sciences and Education (2012) *Using Science as Evidence in Public Policy* (eds K. Prewitt, T.A. Schwandt, and M.L. Straf), Washington, DC: The National Academies Press.

Nilsen, E. (2022) "Supreme Court hears case that could limit EPA's authority to regulate planet-warming emissions from power plants," *CNN Politics*, February 28. Available at: www.cnn.com/2022/02/28/politics/supreme-court-hears-epa-power-plant-emissions-case-climate/index.html (Accessed October 26, 2022).

Nunez, S. (2020) *What Are You Going Through*, New York: Penguin Random House.

O'Connor, J. (2020) *Shadowplay: A Novel*, New York: Europa Editions.

Olena, A. (2020) "2020 in scientists' own words," *The Scientist*, December 23. Available at: www.the-scientist.com/news-opin ion/2020-in-scientists-own-words-68307 (Accessed October 26, 2022).

Oreskes, N. (2019) *Why Trust Science?*, Princeton, NJ: Princeton University Press.

Paine, T. (1804) "Of the word religion, and other words of uncertain signification," *The Prospect*, March 3. Available at: www.thomaspaine.org/essays/religion/prospect-papers.html (Accessed June 22, 2022).

Pariser, E. (2011) *The Filter Bubble: How the New Personalized Web Is Changing What We Read and How We Think*, London: Penguin Books.

Pasquale, F. (forthcoming) "Battle of the experts: the strange career of MetaExpertise," in G. Eyal and T. Medvetz (eds) *Oxford Handbook of Expertise and Democratic Politics*, London: Oxford University Press.

Richard, R. (2020) "Dueling chyrons: CNN, Fox News report from alternate universes during Trump's bizarre coronavirus briefing," *Mediaite*, April 13. Available at: www.mediaite.com/news/dueling-chyrons-cnn-fox-news-report-from-alternate-universes-during-trumps-bizarre-coronavirus-briefing/ (Accessed June 22, 2022).

Richards, J. (2010) "When to doubt a scientific 'consensus'," American Enterprise Institute, March 16. Available at: www.aei.org/articles/when-to-doubt-a-scientific-consensus/ (Accessed October 26, 2022).

Riddell, S., Goldie, S., Hill, A., Eagles, D., and Drew, T. (2020) "The effect of temperature on persistence of SARS-CoV-2 on common surfaces," *Virology Journal*, 17(art 145): 1–7.

Satta, M. and Davidson, L. (2019) "Pernicious epistemically justified distrust and public health skepticism," *Bill of Health*, April 6. Available at: https://blog.petrieflom.law.harvard.edu/2019/04/06/pernicious-epistemically-justified-distrust-and-public-health-skepticism/ (Accessed June 22, 2022).

Schama, S. (1987) *The Embarrassment of Riches: An Interpretation of Dutch Culture in the Golden Age*, New York: Alfred K. Knopf.

Schutte, G. (2005) *Groen Van Prinsterer: His Life and Work* (trans H. van Dijk), Neerlandia: Inheritance Publications.

Schwarz, M. and Thompson, M. (1990) *Divided We Stand: Redefining Politics, Technology and Social Choice*, Philadelphia, PA: University of Pennsylvania Press.

The Scientist (2020) "The top retractions of 2020: the Retraction Watch team takes a look at the most important publishing mistakes this year," December 15. Available at: www.the-scientist.com/news-opinion/the-top-retractions-of-2020-68284 (Accessed October 26, 2022).

Shalizi, C. (2017) "Review of Collins and Evans' *Why Democracies Need Science*," in *The Bactra Review: Occasional and Eclectic Book Reviews* (No. 166), June 25. Available at: http://bactra.org/reviews/collins-evans.html(Accessed October 25, 2022).

Sharfstein, J. (2017) "Science and the Trump administration," *The JAMA Forum*, October 10. Available at: https://jamanetwork.com/journals/jama/fullarticle/2656798 (Accessed June 22, 2022).

Skillen J. (2014) *The Good of Politics: A Biblical, Historical, and Contemporary Introduction*, Grand Rapids, MI: Baker Academic.

Smith, S. (2018) *Pagans and Christians in the City: Culture Wars from the Tiber to the Potomac*, Grand Rapids, MI: Wm. B. Eerdmans Publishing.

Spivak, G.C. (1990) "Strategy, identity, writing," in S. Harasym (ed) *The Post-colonial Critic: Interviews, Strategies, Dialogues*, New York: Routledge, pp 35–49.

Stelter, B. (2021) "Fauci and Birx tell interviewers about the 'nonsense' of the Trump years," *CNN BUSINESS*. Available at: www.cnn.com/2021/01/25/media/fauci-brix-trump-reliable-sources/index.html (Accessed June 22, 2022).

Tavernise, S. (2020) "'It hurts': family rifts deepen in Trump years," *The Washington Post*, November 28, p A14.

Taylor, B. (2020) *Real Life*, New York: Riverhead Books.

Thompson, D. (2021) "Hygiene theater is still a huge waste of time," *The Atlantic*, February 8. Available at: www.theatlantic.com/ideas/archive/2021/02/hygiene-theater-still-waste/617939/ (Accessed June 22, 2022).

Tomsick, E. (2020) "The public health demand for revoking non-medical exemptions to compulsory vaccination statutes," *Journal of Law and Health*, 34(1): 129–56.

Toner, P. (2017) "Wittgenstein on forms of life: a short introduction," *E-Logos—Electronic Journal for Philosophy*, 24(1): 13–18.

Tufekci, Z. (2020) "Why telling people they don't need masks backfired (opinion)," *The New York Times*, March 20. Available at: www.nytimes.com/2020/03/17/opinion/coronavirus-face-masks.html?campaign_id=9&emc=edit_nn_20210118&instance _id=26125&nl=the-morning&regi_id=100596401&segment_ id=49584&te=1&user_id=4f435818fe714149705b758320ac43f9 (Accessed June 22, 2022).

Van Dijk, H. (1975) "Foreword," in G. Groen van Prinsterer, *Unbelief and Revolution: A Series of Lectures in History, Lectures VIII and IX* (ed and trans H. van Dijk), Groen van Amsterdam: Prinsterer Fund, pp v–x.

Van Peursen, C.A. (1969) *Ludwig Wittgenstein: An Introduction to His Philosophy* (trans R. Ambler), London: Faber and Faber.

Van Prinsterer, G. (1989 [1868]) *Unbelief and Revolution: A Series of Historical Lectures*, 2nd edn (trans H. Van Dyke), Jordon Station, Ontario: Wedge Publishing.

Vanhaelen, A. (2005) "Iconoclasm and the creation of images in Emanuel de Witte's 'Old Church in Amsterdam,'" *The Art Bulletin*, 87(2): 251.

Wall Street Journal (2021) "To mandate or not to mandate vaccines: the FDA should move faster on final approval to reassure public," July 27. Available at: www.wsj.com/articles/to-mandate-or-not-to-mandate-vaccines-11627425858?page=1 (Accessed June 22, 2022).

Weinel, M. (2019) "Recognizing counterfeit scientific controversies in science policy contexts: a criteria-based approach," in D.S. Caudill, S.N. Conley, M.E. Gorman, and M. Weinel (eds) *The Third Wave in Science and Technology Studies: Future Research Directions on Expertise and Experience*, London: Palgrave Macmillan, pp 53–70.

Winberg, O. (2017) "Insult politics: Donald Trump, right-wing populism, and incendiary language," *European Journal of American Studies*, 12(2): Art 4.

Winch, P. (1970) "Nature and convention," in P. Winch, *Ethics and Action*, London: Routledge and Kegan Paul, pp 50–72.

Winch, P. (1990) *The Idea of a Social Science and Its Relation to Philosophy*, 2nd edn, London: Routledge.

Witte, J., Jr (2008) *The Reformation of Rights: Law, Religion and Human Rights in Early Modern Calvinism*, Cambridge: University of Cambridge Press.

Wittgenstein, L. (1958) *Philosophical Investigations*, 3rd edn (trans G.E.M. Anscombe), New York: Macmillan.

Wittgenstein, L. (1966) *Lectures and Conversations on Aesthetics, Psychology, and Religious Beliefs* (ed C. Barnett), Oxford: Blackwell Publishers.

Wittgenstein, L. (1969) *On Certainty* (eds G.E.M. Anscombe and G.H. von Wright; trans D. Paul and G.E.M. Anscombe), Oxford: Blackwell Publishers.

Wittgenstein, L. (1971) "Remarks on Frazer's 'Golden Bough,'" *The Human World*, 3: 29–30.

Wittgenstein, L. (1980) *Culture and Value* (ed G.H. von Wright; trans P. Winch), Chicago, IL: University of Chicago Press.

Wood, J. (2020) "Creating God," *The New Yorker*, November 9, p 64.

Wynne, B. (1989) "Sheep farming after Chernobyl: a case study in communicating scientific information," *Environment Science and Policy for Sustainable Development*, 31(2): 10–39.

Zylstra, B. (1975) "Introduction," in L. Kalsbeek, *Contours of Christian Philosophy: An Introduction to Herman Dooyeweerd's Thought* (eds B. Zylstra and J. Zylstra), Toronto: Wedge Publishing Foundation, pp 14–33.

Index

References to endnotes show both the
page number and the note number (77n16).

Printed and bound by CPI Group (UK) Ltd, Croydon, CR0 4YY

27/10/2024

14580558-0001